A TREATISE

ON THE

MOTION OF VORTEX RINGS.

AN ESSAY TO WHICH THE ADAMS PRIZE WAS ADJUDGED
IN 1882, IN THE UNIVERSITY OF CAMBRIDGE.

BY

J. J. THOMSON, M. A.

FELLOW AND ASSISTANT LECTURER OF TRINITY COLLEGE, CAMBRIDGE.

British Library Cataloguing-in-Publication Data
A catalogue record for this book is available from
the British Library

PREFACE.

THE subject selected by the Examiners for the Adams Prize for 1882 was

"A general investigation of the action upon each other of two closed vortices in a perfect incompressible fluid."

In this essay, in addition to the set subject, I have discussed some points which are intimately connected with it, and I have endeavoured to apply some of the results to the vortex atom theory of matter.

I have made some alterations in the notation and arrangement since the essay was sent in to the Examiners, in so doing I have received great assistance from Prof. G. H. Darwin, F.R.S. one of the Examiners, who very kindly lent me the notes he had made on the essay. Beyond these I have not made any alterations in the first three parts of the essay: but to the fourth part, which treats of a vortex atom theory of chemical action, I have made some additions in the hope of making the theory more complete: paragraph 60 and parts of paragraphs 58 and 59 have been added since the essay was sent in to the Examiners.

I am very much indebted to Prof. Larmor of Queen's College, Galway, for a careful revision of the proofs and for many valuable suggestions.

J. J. THOMSON.

TRINITY COLLEGE, CAMBRIDGE.
October 1*st*, 1883.

CONTENTS.

CONTENTS.

PART III.

PART IV.

IN this Essay the motion of a fluid in which there are circular vortex rings is discussed. It is divided into four parts, Part I. contains a discussion of the vibrations which a single vortex ring executes when it is slightly disturbed from its circular form. Part II. is an investigation of the action upon each other of two vortex rings which move so as never to approach closer than by a large multiple of the diameter of either; at the end of this section the effect of a sphere on a circular vortex ring passing near it is found. Part III. contains an investigation of the motion of two circular vortex rings linked through each other; the conditions necessary for the existence of such a system are discussed and the time of vibration of the system investigated. It also contains an investigation of the motion of three, four, five, or six vortices arranged in the most symmetrical way, *i.e.* so that any plane perpendicular to their directions cuts their axes in points forming the angular points of a regular polygon; and it is proved that if there are more than six vortices arranged in this way the steady motion is unstable. Part IV. contains some applications of the preceding results to the vortex atom theory of gases, and a sketch of a vortex atom theory of chemical action.

When we have a mass of fluid under the action of no forces, the conditions that must be satisfied are, firstly, that the expressions for the components of the velocity are such as to satisfy the equation of continuity; secondly, that there should be no discontinuity in the pressure; and, thirdly, that if $F(x, y, z, t) = 0$ be the equation to any surface which always consists of the same fluid particles, such as the surface of a solid immersed in a fluid or the surface of a vortex ring, then

$$\frac{dF}{dt} + u\frac{dF}{dx} + v\frac{dF}{dy} + w\frac{dF}{dz} = 0,$$

where the differential coefficients are partial, and u, v, w are the velocity components of the fluid at the point x, y, z. As we use in the following work the expressions given by Helmholtz for the velocity components at any point of a mass of fluid in which there is vortex motion; and as we have only to deal with vortex motion which is distributed throughout a volume and not spread over a surface, there will be no discontinuity in the velocity, and so no discontinuity in the pressure; so that the third is the only condition we have explicitly to consider. Thus our method is very simple. We substitute in the equation

$$\frac{dF}{dt}+u\frac{dF}{dx}+v\frac{dF}{dy}+w\frac{dF}{dz}=0$$

the values of u, v, w given by the Helmholtz equations, and we get differential equations sufficient to solve any of the above problems.

We begin by proving some general expressions for the momentum, moment of momentum, and kinetic energy of a mass of fluid in which there is vortex motion. In equation (9) § 7 we get the following expression for the kinetic energy of a mass of fluid in which the vortex motion is distributed in circular vortex rings,

$$T=\Sigma\left\{2\mathfrak{I}V-\left(f\frac{d\mathfrak{P}}{dt}+g\frac{d\mathfrak{Q}}{dt}+h\frac{d\mathfrak{R}}{dt}\right)\right\}+\tfrac{1}{2}\rho\iint(u^2+v^2+w^2)\,pdS,$$

where T is the kinetic energy; $\mathfrak{I}$ the momentum of a single vortex ring; $\mathfrak{P}$, $\mathfrak{Q}$, $\mathfrak{R}$ the components of this momentum along the axes of x, y, z respectively; V the velocity of the vortex ring; f, g, h the coordinates of its centre; p the perpendicular from the origin on the tangent plane to the surface containing the fluid; and ρ the density of the fluid. When the distance between the rings is large compared with the diameters of the rings, we prove in § 56 that the terms

$$\Sigma\left(f\frac{d\mathfrak{P}}{dt}+g\frac{d\mathfrak{Q}}{dt}+h\frac{d\mathfrak{R}}{dt}\right)$$

for any two rings may be expressed in the following forms:

$$-2m\pi\rho a^2 r\frac{dS}{dr},$$

or

$$\frac{mm'\pi\rho a^2a'^2}{r^3}(3\cos\theta\cos\theta'-\cos\epsilon),$$

where r is the distance between the centres of the rings; m and m' the strengths of the rings, and a and a' their radii; S the velocity due to one vortex ring perpendicular to the plane of the other; ϵ is the angle between their directions of motion; and θ, θ' the angles their directions of motion make with the line joining their centres.

These equations are, I believe, new, and they have an important application in the explanation of Boyle's law (see § 56).

We then go on to consider the vibrations of a single vortex ring disturbed slightly from its circular form; this is necessary for the succeeding investigations, and it possesses much intrinsic interest. The method used is to calculate by the expressions given by Helmholtz the distribution of velocity due to a vortex ring whose central line of vortex core is represented by the equations

$$\rho = a + \Sigma\,(\alpha_n \cos n\psi + \beta_n \sin n\psi),$$
$$z = \mathfrak{z} + \Sigma\,(\gamma_n \cos n\psi + \delta_n \sin n\psi),$$

where ρ, z, and ψ are semi-polar coordinates, the normal to the mean plane of the central line of the vortex ring through its centre being taken as the axis of z and where the quantities α_n, β_n, γ_n, δ_n are small compared with a. The transverse section of the vortex ring is small compared with its aperture. We make use of the fact that the velocity produced by any distribution of vortices is proportional to the magnetic force produced by electric currents coinciding in position with the vortex lines, and such that the strength of the current is proportional to the strength of the vortex at every point. If currents of electricity flow round an anchor ring, whose transverse section is small compared with its aperture, the magnetic effects of the currents are the same as if all the currents were collected into one flowing along the circular axis of the anchor ring (Maxwell's *Electricity and Magnetism*, 2nd ed. vol. II. § 683). Hence the action of a vortex ring of this shape will be the same as one of equal strength condensed at the central line of the vortex core. To calculate the values of the velocity components by Helmholtz's expressions we have to evaluate $\int_0^\pi \frac{\cos n\theta \, . \, d\theta}{\sqrt{(q-\cos\theta)}}$, when q is very nearly unity. This integral occurs in the Planetary Theory in the expansion of the Disturbing Function, and various expressions have been found for it; the case, however, when q is nearly unity is not important in that theory, and no expressions have been given which converge quickly in this case. It was therefore necessary to investigate some expressions for this integral which would converge quickly in this case; the result of this investigation is given in equation 25, viz.

$$\frac{1}{\pi}\int_0^{2\pi} \frac{\cos n\theta \, . \, d\theta}{\sqrt{(q-\cos\theta)}}$$
$$= \frac{\sqrt{2}}{\pi} F(\tfrac{1}{2} - n, \tfrac{1}{2} + n, 1, -\tfrac{1}{2}x) \left\{\log \frac{16\,(q+1)}{q-1} - 4\left(1 + \tfrac{1}{3} + \dots \frac{1}{2n-1}\right)\right\}$$
$$+ \frac{\sqrt{2}}{\pi}\left\{K_1 (n^2 - \tfrac{1}{4}) \frac{x}{2} + K_2 (n^2 - \tfrac{1}{4})(n^2 - \tfrac{9}{4}) \frac{1}{(2\,!)^2} \frac{x^2}{2^2}\right.$$
$$+ K_3 (n^2 - \tfrac{1}{4})(n^2 - \tfrac{9}{4})(n^2 - \tfrac{25}{4}) \frac{1}{(3\,!)^2} \frac{x^3}{2^3} + \dots$$
$$\left. + K_m (n^2 - \tfrac{1}{4})(n^2 - \tfrac{9}{4}) \dots (n^2 - \tfrac{1}{4}(2m-1)^2) \frac{1}{(m\,!)^2} \frac{x^m}{2^m} + \dots \right\},$$

where $K_m = 1 + \frac{1}{3} + \dots \frac{1}{2m-1}$, and $q = 1 + x$; $F(\;)$ denotes as usual the hyper-geometrical series.

In equations 10—18 the expressions for the components of the velocity due to the disturbed vortex at any point in the fluid are given, the expressions going up to and including the squares of the small quantities $\alpha_n, \beta_n, \gamma_n, \delta_n$; from these equations, and the condition that if $F(x, y, z, t) = 0$ be the equation to the surface of a vortex ring, then

$$\frac{dF}{dt} + u\frac{dF}{dx} + v\frac{dF}{dy} + w\frac{dF}{dz} = 0,$$

we get

$$\frac{d\alpha_n}{dt} = -\frac{1}{4}\frac{m\gamma_n}{\pi a^2}n^2\left\{\log\frac{64a^2}{e^2} - 4f(n) - 1\right\}\dots\text{(equation 37)},$$

where m is the strength of the vortex, e the radius of the transverse section, and $f(n) = 1 + \frac{1}{3} + \dots \frac{1}{2n-1}$:

$$\frac{d\mathfrak{z}}{dt} = \frac{m}{2\pi a}\left(\log\frac{8a}{e} - 1\right)\dots\text{(equation 41)},$$

this is the velocity of translation, and this value of it agrees very approximately with the one found by Sir William Thomson:

$$\frac{d\gamma_n}{dt} = \frac{1}{4}\frac{m\alpha_n}{\pi a^2}(n^2-1)\left\{\log\frac{64a^2}{e^2} - 4f(n) - 1\right\}:\ \text{(equation 42)}.$$

We see from this expression that the different parts of the vortex ring move forward with slightly different velocities, and that the velocity of any portion of it is Va/ρ, where V is the undisturbed velocity of the ring, and ρ the radius of curvature of the central line of vortex core at the point under consideration; we might have anticipated this result.

These equations lead to the equation

$$\frac{d^2\alpha_n}{dt^2} + n^2(n^2-1)L^2\alpha_n = 0:\text{(equation 44)},$$

where
$$L = \frac{m}{4\pi a^2}\left\{\log\frac{64a^2}{e^2} - 4f(n) - 1\right\}.$$

Thus we see that the ring executes vibrations in the period

$$\frac{2\pi}{L\sqrt{\{n^2(n^2-1)\}}};$$

thus the circular vortex ring, whose transverse section is small compared with its aperture, is stable for all displacements of its central line of vortex core. Sir William Thomson has proved that it is stable for all small alterations in the shape of its transverse section; hence we conclude that it is stable for all small displacements. A limiting case of the circular vortex ring is the straight columnar vortex column; we find what our expressions for the times of vibration reduce to in this limiting case, and find that they agree very approximately with those found by Sir William Thomson, who has investigated the vibrations of a straight columnar vortex. We thus get a confirmation of the accuracy of the work.

In Part II. we find the action upon each other of two vortex rings which move so as never to approach closer than by a large multiple of the diameter of either. The method used is as follows: let the equations to one of the vortices be

$$\rho = a + \Sigma\,(\alpha_n \cos n\psi + \beta_n \sin n\psi),$$

$$z = \mathfrak{z} + \Sigma\,(\gamma_n \cos n\psi + \delta_n \sin n\psi);$$

then, if $\mathfrak{R}$ be the velocity along the radius, w the velocity perpendicular to the plane of the vortex, we have

$$\mathfrak{R} = \Sigma\left(\frac{d\alpha_n}{dt}\cos n\psi + \frac{d\beta_n}{dt}\sin n\psi\right),$$

$$w = \frac{d\mathfrak{z}}{dt} + \Sigma\left(\frac{d\gamma_n}{dt}\cos n\psi + \frac{d\delta_n}{dt}\sin n\psi\right);$$

and, equating coefficients of $\cos n\psi$ in the expression for $\mathfrak{R}$, we see that $d\alpha_n/dt$ equals the coefficients of $\cos n\psi$ in that expression. Hence we expand $\mathfrak{R}$ and w in the form

$$A\cos\psi + B\sin\psi + A'\cos 2\psi + B'\sin 2\psi + \ldots$$

and express the coefficients A, B, A', B' in terms of the time; and thus get differential equations for α_n, β_n, γ_n, δ_n. The calculation of these coefficients is a laborious process and occupies pp. 38—46. The following is the result of the investigation: If two vortex rings (I.) and (II.) pass each other, the vortex (I.) moving with the velocity p, the vortex (II.) with the velocity q, their directions of motion making an angle ϵ with each other; and if c is the shortest distance between the centres of the vortex rings, $\mathfrak{g}$ the shortest distance between the paths of the vortices, m and

m' the strengths of the vortices (I.) and (II.) respectively, a, b their radii, and k their relative velocity; then if the equation to the plane of the vortex ring (II.), after the vortices have separated so far that they cease to influence each other, be

$$z = \mathfrak{z} + \gamma' \cos\psi + \delta' \sin\psi + \ldots,$$

where the axis of z is the normal to the undisturbed plane of vortex (II.), we have

$$\gamma' = \frac{2ma^2b}{c^4k^4}\sin^2\epsilon \,.\, pq\,(q - p\cos\epsilon)\,\sqrt{(c^2 - \mathfrak{g}^2)}\left(1 - \frac{4\mathfrak{g}^2}{c^2}\right) : \text{(equation 69)},$$

$$\delta' = -\frac{2ma^2b\,\mathfrak{g}\sin^2\epsilon}{c^4k^3}\,pq\left(1 - \frac{4\mathfrak{g}^2}{3c^2}\right) \ldots\ldots\ldots\ldots\ldots\ldots\ldots \text{(equation 71)},$$

and the radius of the ring is increased by

$$\frac{ma^2b\,p^2q}{c^4k^4}\sin^2\epsilon\sqrt{(c^2 - \mathfrak{g}^2)}\left(1 - \frac{4\mathfrak{g}^2}{c^2}\right) . \text{ (equation 74)},$$

where $\sqrt{(c^2 - \mathfrak{g}^2)}$ is positive or negative according as the vortex (II.) does or does not intersect the shortest distance between the paths of the centres of the vortices before the vortex (I.).

The effects of the collision may be divided in three parts: firstly, the effect upon the radii of the vortex rings; secondly, the deflection of their paths in a plane perpendicular to the plane containing parallels to the original directions of motion of the vortices; and, thirdly, the deflection of their paths in the plane parallel to the original directions of motion of both the vortex rings.

Let us first consider the effect upon the radii. Let $\mathfrak{g} = c\cos\phi$, thus ϕ is the angle which the line joining the centres of the vortex rings when they are nearest together makes with the shortest distance between the paths of the centres of the vortex rings; ϕ is positive for the vortex ring which first intersects the shortest distance between the paths negative for the other ring.

The radius of the vortex ring (II.) is diminished by

$$\frac{ma^2b}{c^3k^4}\,p^2q\sin^2\epsilon\sin 3\phi.$$

Thus the radius of the ring is diminished or increased according as $\sin 3\phi$ is positive or negative. Now ϕ is positive for one vortex ring negative for the other, thus $\sin 3\phi$ is positive for one vortex ring negative for the other, so that if the radius of one vortex ring is increased by the collision the radius of the other will be diminished. When ϕ is less than $60°$ the vortex ring which first passes through the shortest distance between the paths of the centres of the rings diminishes in radius and the other one increases. When ϕ is greater than 60° the vortex ring which first passes through the shortest distance between the paths increases in radius and the other one diminishes. When the paths of the centres of the vortex rings intersect ϕ is 90° so that the vortex ring which first passes through the shortest distance, which in this case is the point of intersection of the paths, is the one which increases in radius. When ϕ is zero or the vortex rings intersect the shortest distance simultaneously there is no change in the radius of either vortex ring, and this is also the case when ϕ is 60°.

Let us now consider the bending of the path of the centre of one of the vortex rings perpendicular to the plane which passes through the centre of the other ring and is parallel to the original paths of both the vortex rings.

We see by equation (71) that the path of the centre of the vortex ring (II.) is bent towards this plane through an angle

$$\tfrac{3}{8}\frac{ma^2}{c^3k^3}pq\sin^2\epsilon\cos 3\phi,$$

this does not change sign with ϕ and, whichever vortex first passes through the shortest distance, the deflection is given by the rule that the path of a vortex ring is bent towards or from the plane through the centre of the other vortex and parallel to the original directions of both vortices according as $\cos 3\phi$ is positive or negative, so that if ϕ is less than 30° the path of the vortex is bent towards, and if ϕ be greater than 30°, from this plane. It follows from this expression that if we have a large quantity of vortex rings uniformly distributed they will on the whole repel a vortex ring passing by them.

Let us now consider the bending of the paths of the vortices in the plane parallel to the original paths of both vortex rings. Equation (69) shews that the path of the vortex ring (II.) is bent in this plane through an angle

$$-\frac{2ma^2}{c^3k^4}\sin^2\epsilon\sin 3\phi\, pq\,(q-p\cos\epsilon)$$

towards the direction of motion of the other vortex. Thus the direction of motion of one vortex is bent from or towards the direction of motion of the other according as $\sin 3\phi\,(q-p\cos\epsilon)$ is positive or negative. Comparing this result with the result for the change in the radius, we see that if the velocity of a vortex ring (II.) be greater than the velocity of the other vortex (I.) resolved along the direction of motion of (II.), then the path of each vortex will be bent towards the direction of motion of the other when its radius is increased and away from the direction of motion of the other when its radius is diminished, while if the velocity of the vortex be less than the velocity of the other resolved along its direction of motion, the direction of motion will be bent from the direction of the other when its radius is increased and *vice versâ*. The rules for finding the alteration in the radius were given before.

Equation (75) shews that the effect of the collison is the same as if an impulse

$$-\frac{pq\mathfrak{J}\cdot\mathfrak{J}'}{\pi\rho c^3 k^3}\sin^2\epsilon\sin 3\phi,$$

parallel to the resultant of velocities $p-q\cos\epsilon$, and $q-p\cos\epsilon$ along the paths of vortices (II.) and (I.) respectively and an impulse

$$-\frac{pq\mathfrak{J}\cdot\mathfrak{J}'}{3\pi\rho c^3 k^3}\sin^2\epsilon\cos 3\phi,$$

parallel to the shortest distance between the original paths of the vortex rings, were given to one of the vortices and equal and opposite impulses to the other; here $\mathfrak{J}$ and $\mathfrak{J}'$ are the momenta of the vortices.

We then go on to investigate the other effects of the collision. We find that the collision changes the shapes of the vortices as well as their sizes and directions of motion. If the two vortices are equal and their paths intersect, equations (78) and (79) shew that, after collision, their central lines of vortex core are represented by the equations

$$\rho = a - \frac{m\pi n^4 a^4\sqrt{2}}{4k^5\sqrt{3}}\,\frac{\epsilon^{-nc/k}}{(nc/k)^{\frac{1}{2}}}\sin(2\psi+nt+\epsilon),$$

$$z = \mathfrak{z} + \frac{m\pi n^4 a^4\sqrt{2}}{8k^5}\,\frac{\epsilon^{-nc/k}}{(nc/k)^{\frac{1}{2}}}\cos(2\psi+nt+\epsilon),$$

where $2\pi/n$ is the free period of elliptic vibration of the circular axis. These are the equations to twisted ellipses, whose ellipticities are continually changing; thus the collision sets the vortex ring vibrating about its circular form.

We then go on to consider the changes in size, shape, and direction of motion, which a circular vortex ring suffers when placed in a mass of fluid in which there is a distribution of velocity given by a velocity potential Ω. We prove that if $\frac{d}{dh}$ denotes differentiation along the direction of motion of the vortex ring, l, m, n the direction cosines of this direction of motion, and a the radius of the ring,

$$\left.\begin{aligned}\frac{da}{dt}&=-\tfrac{1}{2}a\frac{d^2\Omega}{dh^2}\\ \frac{dl}{dt}&=l\frac{d^2\Omega}{dh^2}-\frac{d^2\Omega}{dx\,dh}\\ \frac{dm}{dt}&=m\frac{d^2\Omega}{dh^2}-\frac{d^2\Omega}{dy\,dh}\\ \frac{dn}{dt}&=n\frac{d^2\Omega}{dh^2}-\frac{d^2\Omega}{dz\,dh}\end{aligned}\right\}\text{(equation 80).}$$

The first of these equations shews that the radius of a vortex ring placed in a mass of fluid will increase or decrease according as the velocity at the centre of the ring along the straight axis decreases or increases as we travel along a stream line through the centre. We apply these equations to the case of a circular vortex ring moving past a fixed sphere, and find the alteration in the radius and the deflection.

In Part III. we consider vortex rings which are linked through each other. We shew that if the vortex rings are of equal strengths and approximately circular they must both lie on the surface of an anchor ring whose transverse section is small compared with its aperture, the manner of linking being such that there are always portions of the two vortex rings at opposite extremities of a diameter of the transverse section. The two vortex rings rotate with an angular velocity $2m/\pi d^2$ round the circular axis of the anchor ring, whilst this circular axis moves forward with the comparatively slow velocity $\frac{m}{2\pi a}\log\frac{64a^2}{e^2}$, where m is the strength and e the radius of the transverse section of the vortex ring, a is the radius of the circular axis of the anchor ring and d the diameter of its transverse section.

We begin by considering the effect which the proximity of the two vortex rings has upon the shapes of their cross sections; since the distance between the rings is large compared with the radii of their transverse sections and the two rings are always nearly parallel, the problem is very approximately the same as that of two parallel straight columnar vortices, and as the mathematical work is more simple for this case, this is the one we consider. By means of a Lemma (§ 33) which enables us to transfer cylindrical harmonics from one origin to another, we find that the centres of the transverse sections of the vortex columns describe circles with the centre of gravity of the two cross sections of the vortex columns as centre, and that the shapes of their transverse sections keep changing, being always approximately elliptical and oscillating about the circular shape, the ellipticity and time of vibration is given by equation (89). We then go on to discuss the transverse vibrations of the central lines of vortex core of two equal vortex rings linked together. We find that for each mode of deformation there are two periods of vibration, a quick one and a slow one.

If the equations to the central line of one of the vortex rings be

$$\rho = a + \alpha_n \cos n\psi + \beta_n \sin n\psi,$$

$$z = \mathfrak{z} + \gamma_n \cos n\psi + \delta_n \sin n\psi,$$

and the equations to the circular axis of the other be of the same form with α_n', β_n', γ_n', δ_n', written for α_n, β_n, γ_n, δ_n, we prove

$$\left.\begin{aligned} \alpha_n &= A\cos(\nu t + \epsilon) - B\cos(\mu t + \epsilon') \\ \alpha_n' &= A\cos(\nu t + \epsilon) + B\cos(\mu t + \epsilon') \\ \gamma_n &= \frac{\sqrt{(n^2-1)}}{n} A\sin(\nu t + \epsilon) + B\sin(\mu t + \epsilon') \\ \gamma_n' &= \frac{\sqrt{(n^2-1)}}{n} A\sin(\nu t + \epsilon) - B\sin(\mu t + \epsilon') \end{aligned}\right\} \text{(equation 96)},$$

where

$$\nu = \frac{m}{2\pi a^3}\sqrt{\{n^2(n^2-1)\}}\log\frac{64a^2}{de},$$

$$\mu = \frac{m}{\pi}\left\{\frac{2}{d^2} - \frac{(2n^2-1)}{4a^2}\log\frac{d}{e}\right\}.$$

Thus, if the conditions allow of the vortices being arranged in this way the motion is stable. In § 41 we discuss the condition necessary for the existence of such an arrangement of vortex rings; the result is, that if I be the momentum, Γ the resultant moment of momentum, r the number of times the vortices are linked through each other, and ρ the density of the fluid, then I, Γ are constants determining the size of the system, and the conditions are that

$$I = 4m\pi\rho a^2,$$

$$\Gamma = m\pi\rho r a d^2.$$

These equations determine a and d; from these equations we get

$$\frac{d^2}{a^2} = \frac{4\Gamma(4m\pi\rho)^{\frac{1}{2}}}{rI^{\frac{3}{2}}}.$$

Now d^2/a^2 must be small, hence the condition that the rings should be approximately circular and the motion steady and stable, is that $\Gamma(4m\pi\rho)^{\frac{1}{2}}/rI^{\frac{3}{2}}$ should be small. We then go on to consider the case of two unequal vortex rings, and in § (43) we arrive at results similar in character to those we have been describing; the chief difference is that the system cannot exist unless the moment of momentum has a certain value which is given in equation (105), and which only depends on the strengths and volumes of the vortices, and the number of times they are linked through each other.

In the latter half of Part III. we consider the case when n vortices are twisted round each other in such a way that they all lie on the surface of an anchor ring and their central lines of vortex core cut the plane of any transverse section of the anchor ring at the angular point of a regular polygon inscribed in this cross section. We find the times of vibration when n equals 3, 4, 5, or 6, and prove that the motion is unstable for seven or more vortices, so that not more than six vortices can be arranged in this way.

Part IV. contains the application of these results to the vortex atom theory of gases, and to the theory of chemical combination.

ON THE MOTION OF VORTEX RINGS.

§ 1. THE theory that the properties of bodies may be explained by supposing matter to be collections of vortex lines in a perfect fluid filling the universe has made the subject of vortex motion at present the most interesting and important branch of Hydrodynamics. This theory, which was first started by Sir William Thomson, as a consequence of the results obtained by Helmholtz in his epoch-making paper "Ueber Integrale der hydrodynamischen Gleichungen welche den Wirbelbewegungen entsprechen" has *à priori* very strong recommendations in its favour. For the vortex ring obviously possesses many of the qualities essential to a molecule that has to be the basis of a dynamical theory of gases. It is indestructible and indivisible; the strength of the vortex ring and the volume of liquid composing it remain for ever unaltered; and if any vortex ring be knotted, or if two vortex rings be linked together in any way, they will retain for ever the same kind of be-knottedness or linking. These properties seem to furnish us with good materials for explaining the permanent properties of the molecule. Again, the vortex ring, when free from the influence of other vortices, moves rapidly forward in a straight line; it can possess, in virtue of its motion of translation, kinetic energy; it can also vibrate about its circular form, and in this way possess internal energy, and thus it affords us promising materials for explaining the phenomena of heat and radiation.

This theory cannot be said to explain what matter is, since it postulates the existence of a fluid possessing inertia; but it proposes to explain by means of the laws of Hydrodynamics all the properties of bodies as consequences of the motion of this fluid. It is thus evidently of a very much more fundamental character than any theory hitherto started; it does not, for example, like the ordinary kinetic theory of gases, assume that the atoms attract each other with a force which varies as that power of the distance which is most convenient, nor can it hope to explain any property of bodies by giving the same property to the atom. Since this theory is the only one that attempts to give any account of the mechanism of the intermolecular forces, it enables us to form much the clearest mental representation of what goes on when one atom influences another. Though the theory is not sufficiently developed for us to say whether or not it succeeds in explaining all the properties of bodies, yet, since it gives to the subject of vortex motion the greater part of the interest it possesses, I shall not scruple to examine the consequences according to this theory of any results I may obtain.

The present essay is divided into four parts: the first part, which is a necessary preliminary to the others, treats of some general propositions in vortex motion and considers at some length the theory of the single vortex ring; the second part treats of the mutual action of two vortex rings which never approach closer than a large multiple of the diameter of either, it also treats of the effect of a solid body immersed in the fluid on a vortex ring passing near it; the third part treats of knotted and linked vortices; and the fourth part contains a sketch of a vortex theory of chemical combination, and the application of the results obtaining in the preceding parts to the vortex ring theory of gases.

It will be seen that the work is almost entirely kinematical; we start with the fact that the vortex ring always consists of the same particles of fluid (the proof of which, however, requires dynamical considerations), and we find that the rest of the work is kinematical. This is further evidence that the vortex theory of matter is of a much more fundamental character than the ordinary solid particle theory, since the mutual action of two vortex rings can be found by kinematical principles, whilst the "clash of atoms" in the ordinary theory introduces us to forces which themselves demand a theory to explain them.

PART I.

Some General Propositions in Vortex Motion.

§ 2. We shall, for convenience of reference, begin by quoting the formulae we shall require. We shall always denote the components of the velocity at the point (x, y, z) of the incompressible fluid by the letters, u, v, w; the components of the angular velocity of molecular rotation will be denoted by ξ, η, ζ.

Velocity.

§ 3. The elements of velocity arising from rotations ξ', η', ζ' in the element of fluid $dx'dy'dz'$ are given by

$$\left.\begin{aligned} \delta u &= \frac{1}{2\pi r^3}\{\eta'(z-z') - \zeta'(y-y')\}\,dx'dy'dz' \\ \delta v &= \frac{1}{2\pi r^3}\{\zeta'(x-x') - \xi'(z-z')\}\,dx'dy'dz' \\ \delta w &= \frac{1}{2\pi r^3}\{\xi'(y-y') - \eta'(x-x')\}\,dx'dy'dz' \end{aligned}\right\}\ldots(1),$$

where r is the distance between the points (x, y, z) and (x', y', z').

Momentum.

§ 4. The value of the momentum may be got by the following considerations: Consider a single closed ring of strength m, the velocity potential at any point in the irrotationally moving fluid due to it is $-\frac{m}{2\pi}$ times the solid angle subtended by the vortex ring at that point, thus it is a many-valued function whose cyclic constant is $2m$. If we close the opening of the ring by a barrier, we shall render the region acyclic. Now we know that the motion at any instant can be generated by applying an impulsive pressure to the surface of the vortex ring and an impulsive pressure over the barrier equal per unit of area to ρ times the cyclic constant, ρ being the density of the fluid. Now if the transverse dimensions of the vortex ring be small in comparison with its aperture, the impulse over it may be neglected in comparison with that over the barrier, and thus we see that the motion can be generated by a normal impulsive pressure over the barrier equal per unit of area to $2m\rho$.

Resolving the impulse parallel to the axis of x, we get

momentum of the whole fluid system parallel to $x = 2m\rho x$
(projection of area of vortex ring on plane yz),

with similar expressions for the components parallel to the axes of y and z.

Thus for a single circular vortex ring, if a be its radius and λ, μ, ν the direction-cosines of the normal to its plane, the components of momentum parallel to the axes of x, y, z respectively are

$$2\pi\rho m a^2\lambda,$$
$$2\pi\rho m a^2\mu,$$
$$2\pi\rho m a^2\nu.$$

The momentum may also be investigated analytically in the following way:

Let P be the x component of the whole momentum of the fluid which moves irrotationally due to a single vortex ring of strength m.

Let Ω be the velocity potential, then

$$P = \iiint \rho \frac{d\Omega}{dx}\, dx\, dy\, dz.$$

Integrating with respect to x,

$$P = \iint \rho\, (\Omega_1 - \Omega_2)\, dy\, dz,$$

where Ω_1 and Ω_2 are the values of Ω at two points on opposite sides of the barrier and infinitely close to it. Now the solid angle subtended by the ring increases by 4π on crossing the boundary, thus

$$\Omega_1 - \Omega_2 = 2m;$$

therefore $$P = 2m \iint \rho\, dy\, dz,$$

where the integration is to be taken all over the barrier closing the vortex ring; if λ, μ, ν be the direction-cosines of the normal to this barrier at any point

$$P = 2m\rho \iint \lambda dS,$$

where dS is an element of the barrier.

Now $$\iint \lambda dS = \tfrac{1}{2}\int\left(y\frac{dz}{ds} - z\frac{dy}{ds}\right)ds,$$

where ds is an element of the boundary of the barrier, *i.e.* an element of the vortex ring, thus

$$P = m\rho\int\left(y\frac{dz}{ds} - z\frac{dy}{ds}\right)ds$$
$$= \rho\iiint(y\zeta - z\eta)\,dx\,dy\,dz,$$

and if we extend the integration over all places where there is vortex motion, this will be the expression for the x component of the momentum due to any distribution of vortex motion.

Thus, if P, Q, R be the components of the momentum along x, y, z respectively,

$$\left.\begin{aligned} P &= \rho\iiint(y\zeta - z\eta)\,dx\,dy\,dz \\ Q &= \rho\iiint(z\xi - x\zeta)\,dx\,dy\,dz \\ R &= \rho\iiint(x\eta - y\xi)\,dx\,dy\,dz \end{aligned}\right\} \text{..................(2).}$$

Again $$\frac{dP}{dt} = \rho\iiint\frac{du}{dt}\,dx\,dy\,dz.$$

But where a force potential V exists,

$$\frac{du}{dt} = 2v\zeta - 2w\eta - \frac{d\chi}{dx},$$

where $$\chi = \int\frac{dp}{\rho} + V + \tfrac{1}{2}\,(\text{vel.})^2$$

(Lamb's *Treatise on the Motion of Fluids*, p. 241); therefore

$$\frac{dP}{dt} = \rho\iiint\left(2v\zeta - 2w\eta - \frac{d\chi}{dx}\right)dx\,dy\,dz.$$

Since χ is single-valued and vanishes at an infinite distance,

$$\iiint\frac{d\chi}{dx}\,dx\,dy\,dz = 0.$$

Again, $$\iiint(v\zeta - w\eta)\,dx\,dy\,dz = 0$$

(Lamb's *Treatise*, p. 161, equation 31); therefore

$$\frac{dP}{dt} = 0,$$

or P is constant. We may prove in a similar way that both Q and R are constant; thus the resultant momentum arising from any distribution of vortices in an unlimited mass of fluid remains constant both in magnitude and direction.

Moment of Momentum.

§ 5. Let L, M, N be the components of the moment of momentum about the axes of x, y, z respectively; let the other notation be the same as before; then for a single vortex ring

$$L = \rho \iiint (wy - vz)\, dx\, dy\, dz$$

$$= \rho \iiint \left(y \frac{d\Omega}{dz} - z \frac{d\Omega}{dy} \right) dx\, dy\, dz$$

$$= \rho \iint \{ y (\Omega_1 - \Omega_2)\, dx\, dy - z (\Omega_1 - \Omega_2)\, dx\, dz \}$$

$$= 2m\rho \iint (z\mu - y\nu)\, dS;$$

this surface integral is, by Stokes' theorem, equal to the line integral

$$\tfrac{1}{2} \int (z^2 + y^2) \frac{dx}{ds}\, ds.$$

So
$$L = m\rho \int (z^2 + y^2) \frac{dx}{ds}\, ds$$

$$= \rho \iiint (z^2 + y^2)\, \xi\, dx\, dy\, dz;$$

and if we extend the integration over all places where there is vortex motion, this will be the expression for the x component of the moment of momentum due to any distribution of vortices.

Thus

$$\left.\begin{aligned} L &= \rho \iiint (y^2 + z^2)\, \xi\, dx\, dy\, dz \\ M &= \rho \iiint (z^2 + x^2)\, \eta\, dx\, dy\, dz \\ N &= \rho \iiint (x^2 + y^2)\, \zeta\, dx\, dy\, dz \end{aligned}\right\} \quad \text{..............(3).}$$

Again,
$$\frac{dL}{dt} = \iiint \left(y \frac{dw}{dt} - z \frac{dv}{dt} \right) dx\, dy\, dz;$$

as before,
$$\frac{dv}{dt} = 2w\xi - 2u\zeta - \frac{d\chi}{dy},$$

$$\frac{dw}{dt} = 2u\eta - 2v\xi - \frac{d\chi}{dz};$$

thus
$$\frac{dL}{dt} = 2 \iiint \{ y (u\eta - v\xi) - z (w\xi - u\zeta) \}\, dx\, dy\, dz$$

$$+ \iiint \left(z \frac{d\chi}{dy} - y \frac{d\chi}{dz} \right) dx\, dy\, dz.$$

Since χ is a single-valued function, the last term vanishes, and

$$\iiint z (w\xi - u\zeta)\, dx\, dy\, dz = \iiint z \left\{ w \left(\frac{dw}{dy} - \frac{dv}{dz} \right) - u \left(\frac{dv}{dx} - \frac{du}{dy} \right) \right\} dx\, dy\, dz.$$

Integrating this by parts, it

$$=\iint(zw^2dxdz - zwv\,dxdy - zuv\,dydz + zu^2dxdz)$$

$$-\iiint\left(zw\frac{dw}{dy} - zv\frac{dw}{dz} - zv\frac{du}{dx} + zu\frac{du}{dy} - vw\right)dxdydz.$$

The surface integrals are taken over a surface at an infinite distance R from the origin; now we know that at an infinite distance u, v, w are at most of the order $\frac{1}{R^3}$, while the element of surface is of the order R^2, and z is of the order R; thus the surface integral is of the order $\frac{1}{R}$ at most, and so vanishes when R is indefinitely great.

Integrating by parts, similar considerations will shew that

$$\iiint zw\frac{dw}{dy}dxdydz = 0,$$

$$\iiint zu\frac{du}{dy}dxdydz = 0;$$

so the integral we are considering becomes

$$\iiint\left(zv\frac{dw}{dz} + zv\frac{du}{dx} + vw\right)dxdydz;$$

or, since $$\frac{du}{dx} + \frac{dv}{dy} + \frac{dw}{dz} = 0,$$

it $$= -\iiint\left(zv\frac{dv}{dy} - vw\right)dxdydz$$

$$= \iiint vw\,dxdydz,$$

since $$\iiint zv\frac{dv}{dy}dxdydz = 0.$$

Similarly $2\iiint y(u\eta - v\xi)\,dxdydz = \iiint vw\,dxdydz,$

and thus $$\frac{dL}{dt} = 2\rho\iiint\{y(u\eta - v\xi) - z(w\xi - u\zeta)\}\,dxdydz = 0;$$

thus L is constant. We may prove in a similar way that M and N are also constant, and thus the resultant moment of momentum arising from any distribution of vortices in an unlimited mass of fluid remains constant both in magnitude and direction. When there are solids in the fluid at a finite distance from the vortices, then the surface integrals do not necessarily vanish, and the momentum and moment of momentum are no longer constant.

Kinetic Energy.

§ 6. The kinetic energy (see Lamb's *Treatise*, § 136)

$$= 2\rho \iiint \{u(y\zeta - z\eta) + v(z\xi - x\zeta) + w(x\eta - y\xi)\} \, dx dy dz;$$

this may be written, using the same notation as before,

$$= 2\rho \Sigma \left[m \int \left\{ u\left(y\frac{dz}{ds} - z\frac{dy}{ds}\right) + v\left(z\frac{dx}{ds} - x\frac{dz}{ds}\right) + w\left(x\frac{dy}{ds} - y\frac{dx}{ds}\right) \right\} ds \right],$$

where Σ means summation for all the vortices.

We shall in subsequent investigations require the expression for the kinetic energy of a system of circular vortex rings. To evaluate the integral for the case of a single vortex ring with any origin O we shall first find its value when the origin is at the centre C'; then we shall find the additional term introduced when we move the origin to a point P on the normal to the plane of the vortex through C', and such that PO is parallel to the plane of the vortex; and, finally, the term introduced by moving the origin from P to O.

When the origin is at C', the integral

$$= 2\rho m \int V a ds,$$

where V is the velocity perpendicular to the plane of the vortex. If V' be the mean value of this quantity taken round the ring, the integral

$$= 4\pi\rho m \, a^2 V'.$$

When we move the origin from C' to P, the additional term introduced

$$= - 2\rho m \int p \, \Re ds,$$

where $\Re$ is the velocity along the radius vector measured outwards, and p the perpendicular from O on the plane of the vortex; thus the integral

$$= - 2m\rho \, p \frac{d}{dt}(\pi a^2).$$

When we change the origin from P to O the additional term introduced

$$= 2\rho m \int c \cos\phi \, V ds,$$

where c is the projection of OC' on the plane of the vortex ring, and ϕ the angle between this projection and the radius vector drawn from the centre of the vortex ring to any point on the circumference.

Let us take as our initial line the intersection of the plane of the vortex ring with the plane through its centre containing the normal and a parallel to the axis of z.

Let ψ be the angle any radius of the vortex ring makes with this initial line, ω the angle which the projection of OC' on the plane of the vortex makes with this initial line; then

$$\phi = \psi - \omega.$$

Let V be expanded in the form

$$V = V' + A\cos\psi + B\sin\psi + C\cos 2\psi + D\sin 2\psi + \&c.,$$

then $$\int \cos\phi \, V ds = \pi a\,(A\cos\omega + B\sin\omega).$$

Since V is not uniform round the vortex ring, the plane of the vortex ring will not move parallel to itself, but will change its aspect. We must express A and B in terms of the rates of change of the direction-cosines of the normal to the plane of the vortex ring.

Let the perpendicular from any point on the vortex ring at the time $t + dt$ on the plane of the ring at the time t be

$$\delta\mathfrak{z} + \delta\alpha\cos\psi + \delta\beta\sin\psi\,;$$

thus the velocity perpendicular to the plane of the vortex

$$= \frac{d\mathfrak{z}}{dt} + \frac{d\alpha}{dt}\cos\psi + \frac{d\beta}{dt}\sin\psi.$$

Comparing this expression with the former expression for the velocity, we get

$$V' = \frac{d\mathfrak{z}}{dt}, \qquad A = \frac{d\alpha}{dt}, \qquad B = \frac{d\beta}{dt}.$$

Fig. 1.

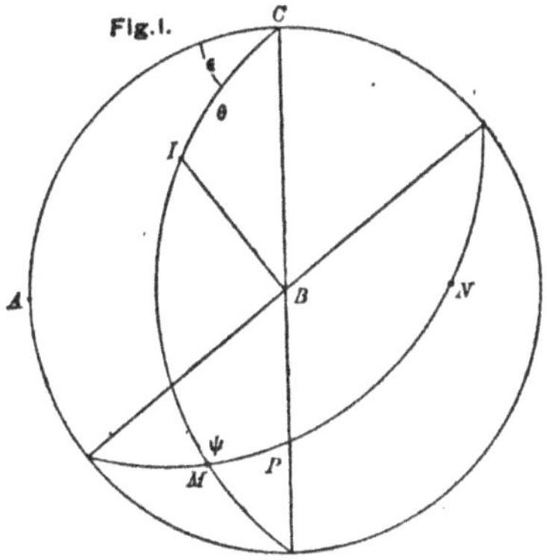

We must now find $\frac{d\alpha}{dt}, \frac{d\beta}{dt}$ in terms of the rates of change of the direction-cosines of the normals to the plane of the ring.

Draw a sphere with its centre at the centre C' of the vortex ring. Let A, B, C be the extremities of axes parallel to the axes x, y, z. Let I be the pole of the ring determined by ϵ and θ as shewn in the figure. Let MN be the ring itself and P any point on it defined by the angle ψ. The displaced position of the plane of the vortex ring may be got by rotating the plane of the ring through an angle $\delta\beta/a$ about the radius vector M in the plane of the ring for which $\psi = 0$, and through an angle $\delta\alpha/a$ about the radius vector N for which $\psi = \frac{\pi}{2}$. The first rotation leaves θ unchanged and diminishes ϵ by $\delta\beta/a \sin\theta$; the second rotation diminishes θ by $\delta\alpha/a$ and leaves ϵ unaltered, thus

$$\delta\theta = -\frac{\delta\alpha}{a},$$

$$\delta\epsilon = -\frac{\delta\beta}{a\sin\theta}.$$

If l, m, n be the direction-cosines of I it is clear that

$$l = \sin\theta\cos\epsilon,\ m = \sin\theta\sin\epsilon,\ n = \cos\theta, \text{ and}$$

$$\therefore\ \delta l = -\frac{\delta\alpha}{a}\cos\theta\cos\epsilon + \frac{\delta\beta}{a}\sin\epsilon,$$

$$\delta m = -\frac{\delta\alpha}{a}\cos\theta\sin\epsilon - \frac{\delta\beta}{a}\cos\epsilon,$$

$$\delta n = \frac{\delta\alpha}{a}\sin\theta.$$

It follows at once that

$$\frac{d\alpha}{dt} = \frac{a}{\sin\theta}\frac{dn}{dt}, \qquad \frac{d\beta}{dt} = a\left(\frac{dl}{dt}\sin\epsilon - \frac{dm}{dt}\cos\epsilon\right),$$

therefore

$$A\cos\omega + B\sin\omega = a\left\{\frac{dn}{dt}\frac{\cos\omega}{\sin\theta} + \left(\frac{dl}{dt}\sin\epsilon - \frac{dm}{dt}\cos\epsilon\right)\sin\omega\right\}.$$

Now if λ', μ', ν' be the direction-cosines of the projection of OC' on the plane of the vortex ring, and f, g, h the coordinates of C',

$$\lambda' = \cos\epsilon\cos\theta\cos\omega - \sin\epsilon\sin\omega,$$

$$\mu' = \sin\epsilon\cos\theta\cos\omega + \cos\epsilon\sin\omega,$$

$$\nu' = -\sin\theta\cos\omega.$$

It is also easily proved that

$$\lambda' = \frac{f - lp}{c},$$

$$\mu' = \frac{g - mp}{c},$$

$$\nu' = \frac{h - np}{c},$$

$$p = lf + mg + nh.$$

So $$\cos\omega = -\frac{\nu'}{\sin\theta} = -\frac{(h - np)}{c\sin\theta} = \frac{lnf + mng - \sin^2\theta \,.\, h}{c\sin\theta},$$

$$\sin\omega = \mu'\cos\epsilon - \lambda'\sin\epsilon = \frac{1}{\sin\theta}(\mu' l - \lambda' m)$$

$$= \frac{1}{c\sin\theta}(lg - mf);$$

thus

$$A\cos\omega + B\sin\omega = \frac{a}{\sin\theta}\left\{\frac{dn}{dt}\cos\omega + \left(\frac{dl}{dt}m - \frac{dm}{dt}l\right)\sin\omega\right\}.$$

This, after substituting for $\cos\omega$ and $\sin\omega$ the values given above,

$$= -\frac{a}{c}\left(h\frac{dn}{dt} + g\frac{dm}{dt} + f\frac{dl}{dt}\right).$$

Thus

$$2\rho mc\pi a\,(A\cos\omega + B\sin\omega) = -2\pi\rho ma^2\left(f\frac{dl}{dt} + g\frac{dm}{dt} + h\frac{dn}{dt}\right).$$

Thus the kinetic energy of the vortex ring

$$= 2\rho m\, 2\pi a^2\, V' - 2m\rho p\frac{d}{dt}(\pi a^2) - 2\pi\rho ma^2\left(f\frac{dl}{dt} + g\frac{dm}{dt} + h\frac{dn}{dt}\right).$$

If I be the momentum of the vortex ring, viz. $2\pi\rho ma^2$, and $\mathfrak{P}$, $\mathfrak{Q}$, $\mathfrak{R}$ the components of I along the axes of x, y, z respectively, this may be written, since $p = lf + mg + nh$,

$$2IV' - \left(f\frac{d\mathfrak{P}}{dt} + g\frac{d\mathfrak{Q}}{dt} + h\frac{d\mathfrak{R}}{dt}\right),$$

and thus the kinetic energy of any system of circular vortex rings

$$= \Sigma\left\{2IV' - \left(f\frac{d\mathfrak{P}}{dt} + g\frac{d\mathfrak{Q}}{dt} + h\frac{d\mathfrak{R}}{dt}\right)\right\} \quad\text{.........(8).}$$

This expression for the kinetic energy is closely analogous to Clausius' expression for the virial in the ordinary molecular theory of gases.

§ 7. We have in the preceding investigation supposed that the bounding surfaces were infinitely distant from the vortices, so that the surface integrals might be neglected; we shall, however, require the expression for the kinetic energy when this is not the case.

The expression

$$2\rho \iiint \{u(y\zeta - z\eta) + v(z\xi - x\zeta) + w(x\eta - y\xi)\}\, dx\,dy\,dz$$

becomes, on integrating by parts and retaining the surface integrals, supposing, however, that the boundaries are fixed so that

$$lu + mv + nw = 0,$$

if l, m, n are the direction-cosines of the normal to the boundary surface,

$$\tfrac{1}{2}\rho \iiint (u^2 + v^2 + w^2)\, dx\,dy\,dz$$
$$- \tfrac{1}{2}\rho \iint (u^2 + v^2 + w^2)(x\,dy\,dz + y\,dx\,dz + z\,dx\,dy),$$

or if dS be an element of the surface and p the perpendicular from the origin on the tangent plane

$$= \tfrac{1}{2}\rho \iiint (u^2 + v^2 + w^2)\, dx\,dy\,dz - \tfrac{1}{2}\rho \iint (u^2 + v^2 + w^2)\, p\,dS.$$

But by the preceding investigation it also equals

$$\Sigma \left\{2IV' - \left(f\frac{d\mathfrak{P}}{dt} + g\frac{d\mathfrak{Q}}{dt} + h\frac{d\mathfrak{R}}{dt}\right)\right\}.$$

Thus T, the kinetic energy, is given by the equation

$$T = \Sigma\left\{2IV' - \left(f\frac{d\mathfrak{P}}{dt} + g\frac{d\mathfrak{Q}}{dt} + h\frac{d\mathfrak{R}}{dt}\right)\right\} + \tfrac{1}{2}\rho \iint (u^2 + v^2 + w^2)\, p\,dS \ldots (9).$$

MOTION OF A SINGLE VORTEX.

§ 8. HAVING investigated these general theorems we shall go on to consider the motion of a single approximately circular vortex ring. We shall suppose that the transverse section of the vortex core is small compared with the aperture of the ring. We know that the velocity produced by any distribution of vortices is proportional to the magnetic force produced by electric currents coinciding in position with the vortex lines, and such that the strength of the current is proportional to the strength of the vortex at every point. Now if currents of electricity flow round an anchor ring, whose transverse section is small compared with its aperture, the magnetic effects of the currents are the same as if all the currents were collected into one flowing along the central line of the anchor ring (Maxwell's *Treatise on Electricity and Magnetism*, 2nd edition, vol. II., § 683). Hence the action of a vortex ring of this shape will be the same as one of equal strength condensed at the central line of the vortex core.

Let the equation to this central line be

$$\rho = a + \alpha_n \cos n\psi + \beta_n \sin n\psi,$$
$$z' = \mathfrak{z} + \gamma_n \cos n\psi + \delta_n \sin n\psi,$$

when z', ρ, ψ are the cylindrical coordinates of a point on the central line of the vortex core, the normal of the vortex ring being taken as the axis of z, the axis of x being the initial line from which the angle ψ is measured. a is the mean radius of the central line of the vortex core, $\mathfrak{z}$ the perpendicular from the origin on the mean plane of the vortex, and α_n, β_n, γ_n, δ_n quantities which are very small compared with a. Let m be the strength of the vortex ring, e the radius of the transverse section of the core. Now, by equations (1), the velocity components due to a vortex of this strength, situated at the central line of the vortex core, are given by

$$u = \frac{m}{2\pi}\int \frac{1}{r^3}\left\{(z-z')\frac{dy'}{ds'} - (y-y')\frac{dz'}{ds'}\right\} ds',$$

$$v = \frac{m}{2\pi}\int \frac{1}{r^3}\left\{(x-x')\frac{dz'}{ds'} - (z-z')\frac{dx'}{ds'}\right\} ds',$$

$$w = \frac{m}{2\pi}\int \frac{1}{r^3}\left\{(y-y')\frac{dx'}{ds'} - (x-x')\frac{dy'}{ds'}\right\} ds',$$

where r is the distance between the points (x, y, z) and (x', y', z'), and the integrals are taken all round the vortex ring.

Now

$$x' = \rho\cos\psi = a\cos\psi + \alpha_n \cos n\psi \cos\psi + \beta_n \sin n\psi \cos\psi,$$

$$y' = \rho\sin\psi = a\sin\psi + \alpha_n \cos n\psi \sin\psi + \beta_n \sin n\psi \sin\psi,$$

therefore

$$\frac{dx'}{d\psi} = -a\sin\psi - \sin\psi\,(\alpha_n \cos n\psi + \beta_n \cos n\psi)$$
$$\qquad - n\cos\psi\,(\alpha_n \sin n\psi - \beta_n \cos n\psi),$$

$$\frac{dy'}{d\psi} = \quad a\cos\psi + \cos\psi\,(\alpha_n \cos n\psi + \beta_n \sin n\psi),$$
$$\qquad - n\sin\psi\,(\alpha_n \sin n\psi - \beta_n \cos n\psi),$$

$$\frac{dz'}{d\psi} = -n\,(\gamma_n \sin n\psi - \delta_n \cos n\psi).$$

In calculating the values of u, v, w we shall retain small quantities up to and including those of the order of the squares of α_n, β_n, γ_n, δ_n. Although, for our present purpose, which is to find the time of oscillation of the vortex about its circular form, we only require to go to the first powers of α_n, &c., yet we go to the higher order of approximation because, when we come to consider the question of knotted vortices, we require the terms containing the squares of these quantities.

If R, ϕ, z be the cylindrical coordinates of the point x, y, z,

$$r^3 = \{\rho^2 + R^2 - 2\rho R\cos(\phi - \psi) + (z-z')^2\}^{\frac{3}{2}},$$

now when we substitute for ρ its value it is evident that $\frac{1}{r^3}$ can be expanded in the form

$$\sum_0^\infty (s)\,(A_s + B_s \cos n\psi + C_s \sin n\psi + D_s \cos 2n\psi + E_s \sin 2n\psi)$$
$$\times \cos s(\psi - \phi),$$

where A_s contains terms independent of $\alpha_n \ldots$, B_s and C_s are of the first, and D_s and E_s of the second order in these quantities.

The part of A_s which is independent of $\alpha_n \dots$ is evidently

$$\frac{1}{\pi}\int_0^{2\pi} \frac{\cos s\theta \, d\theta}{(a^2 + R^2 + z^2 - 2aR\cos\theta)^{\frac{3}{2}}},$$

but we shall investigate the values of all these coefficients later.

Velocity parallel to the axis of x.

§ 9. In the equation

$$u = \frac{m}{2\pi}\int_0^{2\pi} \frac{1}{r^3}\left\{(z - z')\frac{dy'}{d\psi} - (y - y')\frac{dz'}{d\psi}\right) d\psi,$$

the expression to be integrated becomes, when the values for y', z', $\frac{dy'}{ds'}$, $\frac{dz'}{ds'}$ are substituted and the terms arranged in order of magnitude, ζ being written for $z - \mathfrak{z}$,

$$\begin{aligned}\frac{1}{r^3}(\zeta a \cos\psi &+ ny(\gamma_n \sin n\psi - \delta_n \cos n\psi) \\ &+ \tfrac{1}{2}\{(n+1)\zeta\alpha_n + (n-1)a\gamma_n\}\cos(n+1)\psi \\ &- \tfrac{1}{2}\{(n-1)\zeta\alpha_n + (n+1)a\gamma_n\}\cos(n-1)\psi \\ &+ \tfrac{1}{2}\{(n+1)\zeta\beta_n + (n-1)a\delta_n\}\sin(n+1)\psi \\ &- \tfrac{1}{2}\{(n-1)\zeta\beta_n + (n+1)a\delta_n\}\sin(n-1)\psi \\ &+ n(\alpha_n\delta_n - \gamma_n\beta_n)\sin\psi - \tfrac{1}{2}(\alpha_n\gamma_n + \beta_n\delta_n)\cos\psi \\ &- \tfrac{1}{4}(\alpha_n\gamma_n - \beta_n\delta_n)\{\cos(2n+1)\psi + \cos(2n-1)\psi\} \\ &- \tfrac{1}{4}(\alpha_n\delta_n + \gamma_n\beta_n)\{\sin(2n+1)\psi + \sin(2n-1)\psi\}).\end{aligned}$$

Let us consider the term $\frac{m}{2\pi}\int_0^{2\pi} \frac{\zeta a \cos\psi}{r^3} d\psi$.

Expanding $\frac{1}{r^3}$ this equals

$$\begin{aligned}\frac{m}{2\pi}\int_0^{2\pi} d\psi \; \zeta a \cos\psi \, \Sigma_0^\infty(s)\, \{(A_s &+ B_s \cos n\psi + C_s \sin n\psi \\ &+ D_s \cos 2n\psi + E_s \sin 2n\psi)\cos s(\psi - \phi)\}.\end{aligned}$$

Remembering that

$$\int_0^{2\pi} \cos m\psi \cos n\psi \, d\psi = 0 \text{ if } m \text{ does not equal } n,$$

this equals

$$\frac{m}{2\pi}\int_0^{2\pi} d\psi\, \zeta a \cos\psi \times$$

$$+ \quad [\{A_1 \cos(\psi - \phi)]$$

$$+ \tfrac{1}{2}[B_{n+1} \cos\{\psi - (n+1)\phi\} + B_{n-1} \cos\{\psi + (n-1)\phi\}]$$

$$+ \tfrac{1}{2}[C_{n-1} \sin\{\psi + (n-1)\phi\} - C_{n+1} \sin\{\psi - (n+1)\phi\}]$$

$$+ \tfrac{1}{2}[D_{2n+1} \cos\{\psi - (2n+1)\phi\} + D_{2n-1} \cos\{\psi + (2n-1)\phi\}]$$

$$+ \tfrac{1}{2}[E_{2n-1} \sin\{\psi + (2n-1)\phi\} - E_{2n+1} \sin\{\psi - (2n+1)\phi\}])$$

$$= \tfrac{1}{2} ma\zeta [A_1 \cos\phi$$

$$+ \tfrac{1}{2}\{B_{n+1} \cos(n+1)\phi + B_{n-1} \cos(n-1)\phi + C_{n-1} \sin(n-1)\phi + C_{n+1} \sin(n+1)\phi\}$$

$$+ \tfrac{1}{2}\{D_{2n+1} \cos(2n+1)\phi + D_{2n-1} \cos(2n-1)\phi + E_{2n-1} \sin(2n-1)\phi + E_{2n+1} \sin(2n+1)\phi\}].$$

Similarly, we may prove that

$$\frac{m}{2\pi}\int_0^{2\pi} \frac{1}{r^3} ny(\gamma_n \sin n\psi - \delta_n \cos n\psi)\, d\psi$$

$$= \tfrac{1}{2} mny \{A_n(\gamma_n \sin n\phi - \delta_n \cos n\phi) + C_0\gamma_n - B_0\delta_n + \tfrac{1}{2}(B_{2n}\gamma_n - C_{2n}\delta_n) \sin 2n\phi - \tfrac{1}{2}(B_{2n}\delta_n + C_{2n}\gamma_n) \cos 2n\phi\},$$

and that

$$\frac{m}{4\pi}\int_0^{2\pi} \frac{1}{r^3}\{(n+1)\zeta\alpha_n + (n-1)a\gamma_n\} \cos(n+1)\psi\, d\psi$$

$$= \tfrac{1}{4} m \{(n+1)\zeta\alpha_n + (n-1)a\gamma_n\}$$

$$\times \{A_{n+1} \cos(n+1)\phi + \tfrac{1}{2}(B_1 \cos\phi - C_1 \sin\phi + B_{2n+1} \cos(2n+1)\phi + C_{2n+1} \sin(2n+1)\phi)\},$$

and that

$$\frac{m}{4\pi}\int_0^{2\pi} \frac{1}{r^3}\{(n-1)\zeta\alpha_n + (n+1)a\gamma_n\} \cos(n-1)\psi\, d\psi$$

$$= \tfrac{1}{4} m \{(n-1)\zeta\alpha_n + (n+1)a\gamma_n\}$$

$$\times \{A_{n-1} \cos(n-1)\phi + \tfrac{1}{2}(B_1 \cos\phi + C_1 \sin\phi + B_{2n-1} \cos(2n-1)\phi + C_{2n-1} \sin(2n-1)\phi)\},$$

and that

$$\frac{m}{4\pi}\int_0^{2\pi} \frac{1}{r^3}\{(n+1)\zeta\beta_n + (n-1)a\delta_n\} \sin(n+1)\psi\, d\psi$$

$$= \tfrac{1}{4} m \{(n+1)\zeta\beta_n + (n-1)a\delta_n\}$$

$$\times \{A_{n+1} \sin(n+1)\phi + \tfrac{1}{2}(B_1 \sin\phi + C_1 \cos\phi + B_{2n+1} \sin(2n+1)\phi - C_{2n+1} \cos(2n+1)\phi)\},$$

and

$$\frac{m}{4\pi}\int_0^{2\pi}\frac{1}{r^3}\{(n-1)\zeta\beta_n+(n+1)a\delta_n\}\sin(n-1)\psi d\psi$$

$$=\tfrac{1}{4}m\{(n-1)\zeta\beta_n+(n+1)a\delta_n\}$$

$$\times\{A_{n-1}\sin(n-1)\phi+\tfrac{1}{2}(-B_1\sin\phi+C_1\cos\phi+B_{2n-1}\sin(2n-1)\phi$$

$$-C_{2n-1}\cos(2n-1)\phi)\}$$

The integral of the terms involving the products α_n, β_n,...

$$=\tfrac{1}{2}m[nA_1(\alpha_n\delta_n-\beta_n\gamma_n)\sin\phi-\tfrac{1}{2}A_1(\alpha_n\gamma_n+\beta_n\delta_n)\cos\phi$$

$$-\tfrac{1}{4}(\alpha_n\gamma_n-\beta_n\delta_n)\{A_{2n+1}\cos(2n+1)\phi+A_{2n-1}\cos(2n-1)\phi\}$$

$$-\tfrac{1}{4}(\alpha_n\delta_n+\beta_n\gamma_n)\{A_{2n+1}\sin(2n+1)\phi+A_{2n-1}\sin(2n-1)\phi\}].$$

Thus u = terms not containing α_n + terms containing α_n... to the first power + terms containing α_n... to the second power.

The term not containing α_n

$$=\tfrac{1}{2}m\zeta aA_1\cos\phi\text{..................}(10).$$

The terms containing α_n... to the first power

$$=\tfrac{1}{4}m[2nyA_n(\gamma_n\sin n\phi-\delta_n\cos n\phi)$$

$$+\{\zeta aB_{n+1}+[(n+1)\zeta\alpha_n+(n-1)a\gamma_n]A_{n+1}\}\cos(n+1)\phi$$

$$+\{\zeta aB_{n-1}-[(n-1)\zeta\alpha_n+(n+1)a\gamma_n]A_{n-1}\}\cos(n-1)\phi$$

$$+\{\zeta aC_{n+1}+[(n+1)\zeta\beta_n+(n-1)a\delta_n]A_{n+1}\}\sin(n+1)\phi$$

$$+\{\zeta aC_{n-1}-[(n-1)\zeta\beta_n+(n+1)a\delta_n]A_{n-1}\}\sin(n-1)\phi]\quad(11).$$

The terms containing α_n...to the second power

$$=\tfrac{1}{4}m[ny\{\gamma_n(2C_0+B_{2n}\sin 2n\phi-C_{2n}\cos 2n\phi)-\delta_n(2B_0$$

$$+B_{2n}\cos 2n\phi+C_{2n}\sin 2n\phi)\}$$

$$+\{-(\alpha_n\gamma_n+\beta_n\delta_n)A_1+(\zeta\alpha_n-a\gamma_n)B_1+(\zeta\beta_n-a\delta_n)C_1\}\cos\phi$$

$$+n\{2(\alpha_n\delta_n-\beta_n\gamma_n)A_1+(\zeta\beta_n+a\delta_n)B_1-(\zeta\alpha_n+a\gamma_n)C_1\}\sin\phi$$

$$+\{-\tfrac{1}{2}(\alpha_n\gamma_n-\beta_n\delta_n)A_{2n+1}+\tfrac{1}{2}[(n+1)\zeta\alpha_n+(n-1)a\gamma_n]B_{2n+1}$$

$$-\tfrac{1}{2}[(n+1)\zeta\beta_n+(n-1)a\delta_n]C_{2n+1}+a\zeta D_{2n+1}\}\cos(2n+1)\phi$$

$$+\{-\tfrac{1}{2}(\alpha_n\gamma_n-\beta_n\delta_n)A_{2n-1}-\tfrac{1}{2}[(n-1)\zeta\alpha_n+(n+1)a\gamma_n]B_{2n-1}$$

$$+\tfrac{1}{2}[(n-1)\zeta\beta_n+(n+1)a\delta_n]C_{2n-1}+a\zeta D_{2n-1}\}\cos(2n-1)\phi$$

$$+\{-\tfrac{1}{2}(\alpha_n\delta_n+\beta_n\gamma_n)A_{2n+1}+\tfrac{1}{2}[(n+1)\zeta\beta_n+(n-1)a\delta_n]B_{2n+1}$$

$$+\tfrac{1}{2}[(n+1)\zeta\alpha_n+(n-1)a\gamma_n]C_{2n+1}+a\zeta E_{2n+1}\}\sin(2n+1)\phi$$

$$+\{-\tfrac{1}{2}(\alpha_n\delta_n+\beta_n\gamma_n)A_{2n-1}-\tfrac{1}{2}[(n-1)\zeta\beta_n+(n+1)a\delta_n]B_{2n-1}$$

$$-\tfrac{1}{2}[(n-1)\zeta\gamma_n+(n+1)a\gamma_n]C_{2n-1}+a\zeta E_{2n-1}\}\sin(2n-1)\phi]\quad(12)$$

§ 10. $$v = \frac{m}{2\pi}\int_0^{2\pi}\frac{1}{r^3}\left\{(x-x')\frac{dz'}{d\psi}-(z-z')\frac{dx'}{d\psi}\right\}d\psi.$$

The expression to be integrated becomes on substitution

$$\begin{aligned}\frac{1}{r^3}[\zeta a\sin\psi - nx(\gamma_n\sin n\psi-\delta_n\cos n\psi)\\ -\tfrac{1}{2}\{(n+1)\zeta\beta_n+(n-1)a\delta_n\}\cos(n+1)\psi\\ -\tfrac{1}{2}\{(n-1)\zeta\beta_n+(n+1)a\delta_n\}\cos(n-1)\psi\\ +\tfrac{1}{2}\{(n+1)\zeta\alpha_n+(n-1)a\gamma_n\}\sin(n+1)\psi\\ +\tfrac{1}{2}\{(n-1)\zeta\alpha_n+(n+1)a\gamma_n\}\sin(n-1)\psi\\ +n(\beta_n\gamma_n-\alpha_n\delta_n)\cos\psi-\tfrac{1}{2}(\alpha_n\gamma_n+\beta_n\delta_n)\sin\psi\\ -\tfrac{1}{4}(\alpha_n\gamma_n-\beta_n\delta_n)\{\sin(2n+1)\psi-\sin(2n-1)\psi\}\\ -\tfrac{1}{4}(\alpha_n\delta_n+\beta_n\gamma_n)\{\cos(2n-1)\psi-\cos(2n+1)\psi\}].\end{aligned}$$

The term

$$\begin{aligned}&\frac{m}{2\pi}\int_0^{2\pi}\frac{1}{r^3}\zeta a\sin\psi\,d\psi\\ &=\tfrac{1}{2}ma\zeta[aA_1\sin\phi+\tfrac{1}{2}\{B_{n+1}\sin(n+1)\phi-B_{n-1}\sin(n-1)\phi\\ &\qquad -C_{n+1}\cos(n+1)\phi+C_{n-1}\cos(n-1)\phi\}\\ &+\tfrac{1}{2}\{D_{2n+1}\sin(2n+1)\phi-D_{2n-1}\sin(2n-1)\phi-E_{2n+1}\cos(2n+1)\phi\\ &\qquad +E_{2n-1}\cos(2n-1)\phi\}].\end{aligned}$$

The term

$$\begin{aligned}&-\frac{m}{2\pi}\int_0^{2\pi}\frac{nx}{r^3}(\gamma_n\sin n\psi-\delta_n\cos n\psi)\,d\psi\\ &\quad=-\tfrac{1}{2}mnx\{A_n(\gamma_n\sin n\phi-\delta_n\cos n\phi)+C_0\gamma_n-\beta_0\delta_n\\ &\qquad+\tfrac{1}{2}(B_{2n}\gamma_n-C_{2n}\delta_n)\sin 2n\phi\\ &\qquad-\tfrac{1}{2}(B_{2n}\delta_n+C_{2n}\gamma_n)\cos 2n\phi\}.\end{aligned}$$

The term

$$\begin{aligned}&-\frac{m}{4\pi}\{(n+1)\zeta\beta_n+(n-1)a\delta_n\}\int_0^{2\pi}\frac{1}{r^3}\cos(n+1)\psi\,.\,d\psi\\ &\quad=-\tfrac{1}{4}m\{(n+1)\zeta\beta_n+(n-1)a\delta_n\}\\ &\times\{A_{n+1}\cos(n+1)\phi+\tfrac{1}{2}(B_1\cos\phi-C_1\sin\phi+B_{2n+1}\cos(2n+1)\phi\\ &\qquad+C_{2n+1}\sin(2n+1)\phi)\}.\end{aligned}$$

The term

$$\begin{aligned}&-\frac{m}{4\pi}\{(n-1)\zeta\beta_n+(n+1)a\delta_n\}\int_0^{2\pi}\frac{1}{r^3}\cos(n-1)\psi\,d\psi\\ &\quad=-\tfrac{1}{4}m\{(n-1)\zeta\beta_n+(n+1)a\delta_n\}\\ &\times\{A_{n-1}\cos(n-1)\phi+\tfrac{1}{2}(B_1\cos\phi+C_1\sin\phi+B_{2n-1}\cos(2n-1)\phi\\ &\qquad+C_{2n-1}\sin(2n-1)\phi)\}\end{aligned}$$

The term

$$\frac{m}{4\pi}\{(n+1)\zeta\alpha_n+(n-1)a\gamma_n\}\int_0^{2\pi}\frac{1}{r^3}\sin(n+1)\psi\,d\psi$$

$$=\tfrac{1}{4}m\{(n+1)\zeta\alpha_n+(n-1)a\gamma_n\}$$

$$\times\{A_{n+1}\sin(n+1)\phi+\tfrac{1}{2}(B_1\sin\phi+C_1\cos\phi+B_{2n+1}\sin(2n+1)\phi$$
$$-C_{2n+1}\cos(2n+1)\phi)\}$$

The term

$$\frac{m}{4\pi}\{(n-1)\zeta\alpha_n+(n+1)a\gamma_n\}\int_0^{2\pi}\frac{1}{r^3}\sin(n-1)\psi\,d\psi$$

$$=\tfrac{1}{4}m\{(n-1)\zeta\alpha_n+(n+1)a\gamma_n\}$$

$$\times[A_{n-1}\sin(n-1)\phi+\tfrac{1}{2}\{-B_1\sin\phi+C_1\cos\phi+B_{2n-1}\sin(2n-1)\phi$$
$$-C_{2n-1}\cos(2n-1)\phi\}]$$

The integral of the terms involving the products α_n, β_n...

$$=\tfrac{1}{2}m[n(\beta_n\gamma_n-\alpha_n\delta_n)A_1\cos\phi-\tfrac{1}{2}(\alpha_n\beta_n+\beta_n\delta_n)A_1\sin\phi$$
$$-\tfrac{1}{4}(\alpha_n\gamma_n-\beta_n\delta_n)\{A_{2n+1}\sin(2n+1)\phi-A_{2n-1}\sin(2n-1)\phi\}$$
$$-\tfrac{1}{4}(\alpha_n\delta_n+\beta_n\gamma_n)\{A_{2n-1}\cos(2n-1)\phi-A_{2n+1}\cos(2n+1)\phi\}].$$

Thus $v=$ terms not containing α_n... + terms containing α_n...to the first power + terms containing α_n... to the second power.

The term not containing $\alpha_n... = \tfrac{1}{2}m\zeta aA_1\sin\phi$ ....(13).

The terms containing α_n... to the first power

$$=\tfrac{1}{4}m[-2nxA_n(\gamma_n\sin n\phi-\delta_n\cos n\phi)$$
$$-\{[(n+1)\zeta\beta_n+(n-1)a\delta_n]A_{n+1}+a\zeta C_{n+1}\}\cos(n+1)\phi$$
$$-\{[(n-1)\zeta\beta_n+(n+1)a\delta_n]A_{n-1}-a\zeta C_{n-1}\}\cos(n-1)\phi$$
$$+\{[(n+1)\zeta\alpha_n+(n-1)a\gamma_n]A_{n+1}+a\zeta B_{n+1}\}\sin(n+1)\phi$$
$$+\{[(n-1)\zeta\alpha_n+(n+1)a\gamma_n]A_{n-1}-a\zeta B_{n-1}\}\sin(n-1)\phi] \quad ...(14).$$

The terms containing α_n...to the second power

$$=\tfrac{1}{4}m[-nx\{\gamma_n(2C_0+B_{2n}\sin 2n\phi-C_{2n}\cos 2n\phi)$$
$$-\delta_n(2B_0+B_{2n}\cos 2n\phi+C_{2n}\sin 2n\phi)\}$$
$$+n\{2(\beta_n\gamma_n-\alpha_n\delta_n)A_1-(\zeta\beta_n+a\delta_n)B_1+(\zeta\alpha_n+a\gamma_n)C_1\}\cos\phi$$
$$-\{-(\alpha_n\gamma_n+\beta_n\delta_n)A_1+(\zeta\alpha_n-a\gamma_n)B_1+(\zeta\beta_n-a\delta_n)C_1\}\sin\phi$$
$$+\{\tfrac{1}{2}(\alpha_n\delta_n+\beta_n\gamma_n)A_{2n+1}-\tfrac{1}{2}[(n+1)\zeta\beta_n+(n-1)a\delta_n]B_{2n+1}$$
$$-\tfrac{1}{2}[(n+1)\zeta\alpha_n+(n-1)a\delta_n]C_{2n+1}-a\zeta E_{2n+1}\}\cos(2n+1)\phi$$
$$+\{-\tfrac{1}{2}(\alpha_n\delta_n+\beta_n\gamma_n)A_{2n-1}-\tfrac{1}{2}[(n-1)\zeta\beta_n+(n+1)a\delta_n]B_{2n-1}$$
$$-\tfrac{1}{2}[(n-1)\zeta\alpha_n+(n+1)a\gamma_n]C_{2n-1}+a\zeta E_{2n-1}\}\cos(2n-1)\phi$$

$$+\{-\tfrac{1}{2}(\alpha_n\gamma_n-\beta_n\delta_n)A_{2n+1}+\tfrac{1}{2}[(n+1)\zeta\alpha_n+(n-1)a\gamma_n]B_{2n+1}$$
$$-\tfrac{1}{2}[(n+1)\zeta\beta_n+(n-1)a\delta_n]C_{2n+1}+a\zeta D_{2n+1}\}\sin(2n+1)\phi$$
$$+\{\tfrac{1}{2}(\alpha_n\gamma_n-\beta_n\delta_n)A_{2n-1}+\tfrac{1}{2}[(n-1)\zeta\alpha_n+(n+1)a\gamma_n]B_{2n-1}$$
$$-\tfrac{1}{2}[(n-1)\zeta\beta_n+(n+1)a\delta_n]C_{2n-1}-a\zeta D_{2n-1}\}\sin(2n+1)\phi\ldots(15)$$

§ 11. $$w=\frac{m}{2\pi}\int_0^{2\pi}\frac{1}{r^3}\left\{(y-y')\frac{dx'}{d\psi}-(x-x')\frac{dy'}{d\psi}\right\}d\psi.$$

The expression to be integrated becomes after substitution

$$\frac{1}{r^3}[a^2-a(y\sin\psi+\cos\psi)+2a(\alpha_n\cos n\psi+\beta_n\sin n\psi)$$
$$+\tfrac{1}{2}(n+1)(y\beta_n-x\alpha_n)\cos(n+1)\psi$$
$$+\tfrac{1}{2}(n-1)(x\alpha_n+y\beta_n)\cos(n-1)\psi$$
$$-\tfrac{1}{2}(n+1)(y\alpha_n+x\beta_n)\sin(n+1)\psi$$
$$-\tfrac{1}{2}(n-1)(y\alpha_n-x\beta_n)\sin(n-1)\psi$$
$$+(\alpha_n^2\cos^2 n\psi+2\alpha_n\beta_n\cos n\psi\sin n\psi+\beta_n^2\sin^2 n\psi)].$$

The term $\frac{m}{2\pi}\int_0^{2\pi}\frac{a^2}{r^3}d\psi$

$$=\tfrac{1}{2}ma^2(2A_0+B_n\cos n\phi+C_n\sin n\phi+D_{2n}\cos 2n\phi+E_{2n}\sin 2n\phi).$$

The term $-\frac{m}{2\pi}\int_0^{2\pi}\frac{a}{r^3}(y\sin\psi+x\cos\psi)\,d\psi$

putting $x=R\cos\phi$, $y=R\sin\phi$ becomes

$$-\frac{maR}{2\pi}\int_0^{2\pi}\frac{1}{r^3}\cos(\phi-\psi)\,d\psi$$
$$=-\tfrac{1}{4}maR[2A_1+(B_{n+1}+B_{n-1})\cos n\phi+(C_{n+1}+C_{n-1})\sin n\phi$$
$$+(D_{2n+1}+D_{2n-1})\cos 2n\phi+(E_{2n+1}+E_{2n-1})\sin 2n\phi]$$

The term $\frac{m}{\pi}\int_0^{2\pi}\frac{a}{r^3}(\alpha_n\cos n\psi+B_n\sin n\psi)\,d\psi$

$$=ma[A_n(\alpha_n\cos n\phi+\beta_n\sin n\phi)+B_0\alpha_n+C_0\beta_n$$
$$+\tfrac{1}{2}(B_{2n}\alpha_n-C_{2n}\beta_n)\cos 2n\phi+\tfrac{1}{2}(C_{2n}\alpha_n+B_{2n}\beta_n)\sin 2n\phi]$$

The term $\frac{m}{4\pi}(n+1)(y\beta_n-x\alpha_n)\int_0^{2\pi}\frac{1}{r^3}\cos(n+1)\psi\,d\psi$

$$=\frac{m}{4}(n+1)(y\beta_n-x\alpha_n)$$
$$\times\{A_{n+1}\cos(n+1)\phi+\tfrac{1}{2}(B_1\cos\phi-C_1\sin\phi$$
$$+B_{2n+1}\cos(2n+1)\phi+C_{2n+1}\sin(2n+1)\phi)\}.$$

The term $\frac{m}{4\pi}(n-1)(x\alpha_n + y\beta_n)\int_0^{2\pi}\frac{1}{r^3}\cos(n-1)\psi d\psi$

$$= \frac{m}{4}(n-1)(x\alpha_n + y\beta_n)$$
$$\times \{A_{n-1}\cos(n-1)\phi + \tfrac{1}{2}(B_1\cos\phi + C_1\sin\phi + B_{2n-1}\cos(2n-1)\phi + C_{2n-1}\sin(2n-1)\phi)\}$$

The term $-\frac{m}{4\pi}(n+1)(y\alpha_n + x\beta_n)\int_0^{2\pi}\frac{1}{r^3}\sin(n+1)\psi d\psi$

$$= -\frac{m}{4}(n+1)(y\alpha_n + x\beta_n)$$
$$\times \{A_{n+1}\sin(n+1)\phi + \tfrac{1}{2}[B_1\sin\phi + C_1\cos\phi + B_{2n+1}\sin(2n+1)\phi - C_{2n+1}\cos(2n+1)\phi]\}.$$

The term $-\frac{m}{4\pi}(n-1)(y\alpha_n - x\beta_n)\int_0^{2\pi}\frac{1}{r^3}\sin(n-1)\psi d\psi$

$$= -\frac{m}{4}(n-1)(y\alpha_n - x\beta_n)$$
$$\times \{A_{n-1}\sin(n-1)\phi + \tfrac{1}{2}(-B_1\sin\phi + C_1\cos\phi + B_{2n-1}\sin(2n-1)\phi - C_{2n-1}\cos(2n-1)\phi)\}.$$

The term containing the second powers of $\alpha_n \ldots$

$$= \tfrac{1}{2}m\{(\alpha^2_n + \beta^2_n)A_0 + \tfrac{1}{2}(\alpha^2_n - \beta^2_n)A_{2n}\cos 2n\phi + \alpha_n\beta_n A_{2n}\sin 2n\phi\}.$$

Thus w = terms not involving α_n + terms containing $\alpha_n \ldots$ to the first power + terms containing $\alpha_n \ldots$ to the second power.

The terms not involving α_n

$$= \tfrac{1}{2}m(2a^2A_0 - aRA_1) \ldots\ldots\ldots\ldots\ldots\ldots (16).$$

The terms involving $\alpha_n \ldots$ to the first power become after substituting for x and y, $R\cos\phi$ and $R\sin\phi$ respectively

$$\tfrac{1}{2}m[(a^2B_n - \tfrac{1}{2}aR(B_{n+1} + B_{n-1}) + 2a\alpha_n A_n + \tfrac{1}{2}R\alpha_n\{(n-1)A_{n-1} - (n+1)A_{n+1}\})\cos n\phi$$
$$+ (a^2C_n - \tfrac{1}{2}aR(C_{n+1} + C_{n-1}) + 2a\beta_n A_n + \tfrac{1}{2}R\beta_n\{(n-1)A_{n-1} - (n+1)A_{n-1}\})\sin n\phi] \ldots\ldots (17).$$

The term involving $\alpha_n \ldots$ to the second power

$$= m[a\alpha_n B_0 + a\beta_n C_0 - \tfrac{1}{4}\alpha_n B_1 - \tfrac{1}{4}\beta_n C_1 + \tfrac{1}{2}(\alpha^2_n + \beta^2_n)A_0$$
$$+ \tfrac{1}{8}\{R\alpha_n[(n-1)B_{2n-1} - (n+1)B_{2n+1}]$$
$$- R\beta_n[(n-1)C_{2n-1} - (n+1)C_{2n+1}] + 4a(B_{2n}\alpha_n - C_{2n}\beta_n)$$
$$+ 4a^2D_{2n} - 2aR(D_{2n+1} + D_{2n-1}) + 2(\alpha^2_n - \beta^2_n)A_{2n}\}\cos 2n\phi$$

$$+\tfrac{1}{8}\{R\alpha_n[(n-1)C_{2n-1}-(n+1)C_{2n+1}]$$
$$-R\beta_n[(n+1)B_{2n+1}-(n-1)B_{2n-1}]+4a(C_{2n}\alpha_n+B_{2n}\beta_n)$$
$$+4a^2E_{2n}-2aR(E_{2n+1}+E_{2n-1})$$
$$+4\alpha_n\beta_nA_{2n}\}\sin 2n\phi] \quad\ldots\ldots\ldots\ldots(18).$$

§ 12. We must now proceed to determine the values of the quantities which we have denoted by the symbols A_n, B_n, C_n, &c. We have, in fact, to determine the coefficients in the expansion of

$$\frac{1}{\{\rho^2+R^2+\zeta^2-2R\rho\cos(\theta-\phi)\}^{\frac{3}{2}}},$$

or, as it is generally written for symmetry, of

$$\frac{1}{\{1+a^2-2a\cos(\theta-\phi)\}^{\frac{3}{2}}},$$

in the form

$$\mathfrak{A}_0+\mathfrak{A}_1\cos(\theta-\phi)+\ldots\mathfrak{A}_n\cos n(\theta-\phi)+\ldots.$$

This problem also occurs in the Planetary Theory in the expansion of the disturbing function, and consequently these coefficients have received a good deal of attention; they have been considered by, amongst others, Laplace, in the *Mécanique Céleste*, t. I. § 49; Pontecoulant, *Du Système du Monde*, vol. III. chap. II.

These mathematicians obtain series for these coefficients proceeding by ascending powers of a. The case we are most concerned with is when the point whose coordinates are R, z, ϕ is close to the vortex ring, and then R is very nearly equal to ρ and ζ is very small, so that a is very nearly equal to unity, and thus the series given by these mathematicians converge very slowly, and are almost useless for our present purpose. We must investigate some expression which will converge quickly when a is nearly unity.

Our problem in its simplest form may be stated as follows, if

$$\frac{1}{(q-\cos\theta)^{\frac{3}{2}}}=c_0+c_1\cos\theta+\ldots c_n\cos n\theta+\ldots,$$

we have to determine c_n in a form which will converge rapidly if q be nearly unity.

Let
$$\frac{1}{(q-\cos\theta)^{\frac{1}{2}}}=b_0+b_1\cos\theta+\ldots b_n\cos n\theta+\ldots.$$

Then by Fourier's theorem,

$$c_n=\frac{1}{\pi}\int_0^{2\pi}\frac{\cos n\theta}{(q-\cos\theta)^{\frac{3}{2}}}d\theta, \qquad c_0=\frac{1}{2\pi}\int_0^{2\pi}\frac{d\theta}{(q-\cos\theta)^{\frac{3}{2}}},$$

$$b_n=\frac{1}{\pi}\int_0^{2\pi}\frac{\cos n\theta}{(q-\cos\theta)^{\frac{1}{2}}}d\theta, \qquad b_0=\frac{1}{2\pi}\int_0^{2\pi}\frac{d\theta}{(q-\cos\theta)^{\frac{1}{2}}}.$$

Now

$$\frac{d}{d\theta}\frac{\sin n\theta}{(q-\cos\theta)^{\frac{1}{2}}}=\frac{n\cos n\theta}{(q-\cos\theta)^{\frac{1}{2}}}-\tfrac{1}{4}\frac{\{\cos(n-1)\theta-\cos(n+1)\theta\}}{(q-\cos\theta)^{\frac{3}{2}}}\dots(19).$$

Integrating both sides with respect to θ between the limits 0 and 2π, we have

$$0=nb_n-\tfrac{1}{4}(c_{n-1}-c_{n+1}),$$

or

$$4nb_n=c_{n-1}-c_{n+1}\dots\dots(20).$$

Reducing the right-hand side of equation (19) to a common denominator, we have

$$4\frac{d}{d\theta}\frac{\sin n\theta}{(q-\cos\theta)^{\frac{1}{2}}}$$

$$=\frac{4nq\cos n\theta-\{(2n+1)\cos(n-1)\theta+(2n-1)\cos(n+1)\theta\}}{(q-\cos\theta)^{\frac{3}{2}}}.$$

Integrating both sides with respect to θ between the limits 0 and 2π, we get

$$0=4nqc_n-\{(2n+1)c_{n-1}+(2n-1)c_{n+1}\}\dots\dots(21).$$

By means of this and equation (20), we easily get

$$c_n=\frac{2n+1}{(q^2-1)}(qb_n-b_{n+1})\dots\dots(22);$$

and thus, if we know the values of the b's, we can easily get those of the c's, and as the b's are easier to calculate we shall determine them and deduce the values of the c's.

Let $$V=\frac{1}{(q-\cos\theta)^{\frac{1}{2}}}=b_0+b_1\cos\theta+\dots b_n\cos n\theta+\dots.$$

By differentiation we have

$$(1-q^2)\frac{d^2V}{dq^2}-2q\frac{dV}{dq}-\tfrac{1}{4}V=\frac{d^2V}{d\theta^2};$$

hence, substituting for V the value just written and equating the coefficients of $\cos n\theta$ we have

$$(1-q^2)\frac{d^2b_n}{dq^2}-2q\frac{db_n}{dq}+b_n(n^2-\tfrac{1}{4})=0.$$

Let $$b_n=\phi(q)\log\frac{q-1}{16(q+1)}+\psi(q),$$

where $\phi(q)$ and $\psi(q)$ are rational and integral algebraic functions of q.

Substituting in the differential equation, we find

$$(1-q^2)\frac{d^2\phi}{dq^2}-2q\frac{d\phi}{dq}+(n^2-\tfrac{1}{4})\phi=0,$$

$$-4\frac{d\phi}{dq}+(1-q^2)\frac{d^2\psi}{dq^2}-2q\frac{d\psi}{dq}+(n^2-\tfrac{1}{4})\psi=0.$$

Let us change the variable from q to x, where $x = q - 1$, the equations then become

$$x(2+x)\frac{d^2\phi}{dx^2} + 2(1+x)\frac{d\phi}{dx} - (n^2 - \tfrac{1}{4})\phi = 0,$$

$$4\frac{d\phi}{dx} + x(2+x)\frac{d^2\psi}{dx^2} + 2(1+x)\frac{d\psi}{dx} - (n^2 - \tfrac{1}{4})\psi = 0.$$

Let $$\phi = a_0 + a_1 x + \dots a_m x^m + \dots.$$

Substituting in the differential equation for ϕ, we find

$$a_{m+1} = \frac{n^2 - \frac{1}{4} - m \,.\, m+1}{2(m+1)^2} a_m;$$

therefore

$$\phi(x) = a_0 \left\{1 + (n^2 - \tfrac{1}{4})\frac{x}{2}\right.$$

$$\left. + \frac{(n^2 - \frac{1}{4})(n^2 - \frac{9}{4})}{2^2}\left(\frac{x}{2}\right)^2 + \frac{n^2 - \frac{1}{4} \,.\, n^2 - \frac{9}{4} \,.\, n^2 - \frac{25}{4}}{(3\,!)^2}\left(\frac{x}{2}\right)^3 + \dots\right\} \dots (23),$$

or, with the ordinary notation for the hypergeometrical series,

$$\phi(x) = a_0 F(\tfrac{1}{2} - n,\ \tfrac{1}{2} + n,\ 1,\ -\tfrac{1}{2}x).$$

Let $$\psi(x) = \alpha_0 + \alpha_1 x + \alpha_2 x^2 + \dots \alpha_m x^m + \dots.$$

Substituting in the differential equation for $\psi(x)$, we find

$$\alpha_{m+1} = \frac{n^2 - \frac{1}{4} - m \,.\, m+1}{2(m+1)^2}\alpha_m - \frac{2}{m+1} a_{m+1}.$$

So $$\psi(x) = \alpha_0 F(\tfrac{1}{2} - n,\ \tfrac{1}{2} + n,\ 1,\ -\tfrac{1}{2}x)$$

$$- a_0\left\{2(n^2 - \tfrac{1}{4})\frac{x}{2} + 3(n^2 - \tfrac{1}{4})(n^2 - \tfrac{9}{4})\frac{1}{2^2}\frac{x^2}{2^2}\right.$$

$$+ \tfrac{11}{3}(n^2 - \tfrac{1}{4})(n^2 - \tfrac{9}{4})(n^2 - \tfrac{25}{4})\frac{1}{(3\,!)^2}\frac{x^3}{2^3}$$

$$\left. + \tfrac{25}{6}(n^2 - \tfrac{1}{4})(n^2 - \tfrac{9}{4})(n^2 - \tfrac{25}{4})(n^2 - \tfrac{49}{4})\frac{1}{(4\,!)^2}\frac{x^4}{2^4} + \dots\right\} \dots\dots (24),$$

where the general term inside the bracket

$$= 2\left(1 + \tfrac{1}{2} + \dots \frac{1}{m}\right)(n^2 - \tfrac{1}{4})(n^2 - \tfrac{9}{4})\dots(n^2 - \tfrac{1}{4}(2m-1)^2)\frac{1}{(m\,!)^2}\frac{x^m}{2^m}.$$

To complete the solution we have to determine the values of a_0 and α_0. We shall do this by determining the value of b_n when q is very nearly unity, or when x is small.

We may prove, in exactly the same way as we proved equation (20), that

$$4nqb_n = (2n-1)b_{n-1} + (2n+1)b_{n+1},$$

or $$4n(1+x)b_n = (2n-1)b_{n-1} + (2n+1)b_{n+1}.$$

By the help of this sequence equation we can express b_n in terms of b_0 and b_1 in the form

$$b_n = (A + Bx + Cx^2 + Dx^3 + \ldots)\, b_0 + (A' + B'x + C'x^2 + \ldots)\, b_1.$$

We only want now to determine a_0 and α_0, *i.e.* the parts of ψ and ϕ independent of x, thus we only want the coefficients A and A' in the equation just written; now evidently A and A' will be the same as if we put $x = 0$ in the sequence equation and then determine b_n in terms of b_0 and b_1,

The sequence equation becomes, when $x = 0$,

$$4nb_n = (2n-1)\, b_{n-1} + (2n+1)\, b_{n+1},$$

the solution of this is

$$b_n = C + C'\left(1 + \tfrac{1}{3} + \ldots \frac{1}{2n-1}\right),$$

where C and C' are arbitrary constants.

Determining the arbitrary constants in terms of b_0 and b_1, we find

$$b_n = 2b_0 + (b_1 - 2b_0)\left(1 + \tfrac{1}{3} + \ldots \frac{1}{2n-1}\right),$$

for in the sequence equation involving b_0, $2b_0$ must be written instead of b_0.

Now $$2b_0 = \frac{1}{\pi}\int_0^{2\pi} \frac{d\theta}{\sqrt{(q-\cos\theta)}} = \frac{1}{\pi}\int_0^{2\pi} \frac{d\theta}{\sqrt{(q+\cos\theta)}}$$

$$= \frac{1}{\pi}\int_0^{2\pi} \frac{d\theta}{\sqrt{(q+1-2\sin^2\tfrac{1}{2}\theta)}}$$

$$= \frac{1}{\pi\sqrt{(q+1)}}\int_0^{2\pi} \frac{d\theta}{\sqrt{(1-k^2\sin^2\theta)}},$$

where $$k^2 = \frac{2}{q+1},$$

$$= \frac{4}{\pi\sqrt{(q+1)}}\int_0^{\frac{1}{2}\pi} \frac{d\phi}{\sqrt{(1-k^2\sin^2\phi)}}.$$

Now, when k is very nearly unity, we know that

$$\int_0^{\frac{1}{2}\pi} \frac{d\phi}{\sqrt{(1-k^2\sin^2\phi)}} = \log\frac{4}{k_1} \text{ approximately,}$$

where $k_1 = \sqrt{(1-k^2)}$, in our case $= \sqrt{\left(\dfrac{q-1}{q+1}\right)}$.

Therefore, when q is very nearly unity

$$2b_0 = \frac{2\sqrt{2}}{\pi}\log\left\{4\sqrt{\left(\frac{q+1}{q-1}\right)}\right\} \text{ approximately.}$$

$$b_1 = \frac{1}{\pi}\int_0^{2\pi}\frac{\cos\theta\,d\theta}{\sqrt{(q-\cos\theta)}}$$

$$= -\frac{1}{\pi}\int_0^{2\pi}\sqrt{(q-\cos\theta)}\,d\theta + \frac{q}{\pi}\int_0^{2\pi}\frac{d\theta}{\sqrt{(q-\cos\theta)}}$$

$$= -\frac{4}{\pi}\sqrt{(q+1)}\int_0^{\frac{1}{2}\pi}\sqrt{(1-k^2\sin^2\phi)}\,d\phi + 2b_0 q.$$

When k is very nearly unity

$$\int_0^{\frac{1}{2}\pi}\sqrt{(1-k^2\sin^2\phi)}\,d\phi = 1 \text{ approximately;}$$

therefore $$b_1 = -\frac{4\sqrt{2}}{\pi} + 2b_0.$$

Therefore, when q is very nearly unity,

$$b_n = \frac{\sqrt{2}}{\pi}\log\frac{16(q+1)}{q-1} - \frac{4\sqrt{2}}{\pi}\left(1+\tfrac{1}{3}+\dots\frac{1}{2n-1}\right);$$

comparing this with our former solution for b_n, we find

$$a_0 = -\frac{\sqrt{2}}{\pi},$$

and $$a_0 = -\frac{4\sqrt{2}}{\pi}\left(1+\tfrac{1}{3}+\dots\frac{1}{2n-1}\right).$$

Thus $$b_n = \frac{1}{\pi}\int_0^{2\pi}\frac{\cos n\theta}{\sqrt{(q-\cos\theta)}}\,d\theta$$

$$= \frac{\sqrt{2}}{\pi}F(\tfrac{1}{2}-n, \tfrac{1}{2}+n, 1, -\tfrac{1}{2}x)\left\{\log\frac{16(2+x)}{x} - 4\left(1+\tfrac{1}{3}+\dots\frac{1}{2n-1}\right)\right.$$

$$+\frac{\sqrt{2}}{\pi}\left\{K_1(n^2-\tfrac{1}{4})\frac{x}{2} + K_2(n^2-\tfrac{1}{4})(n^2-\tfrac{9}{4})\frac{1}{(2!)^2}\frac{x^2}{2^2}\right.$$

$$+K_3(n^2-\tfrac{1}{4})(n^2-\tfrac{9}{4})(n^2-\tfrac{25}{4})\frac{1}{(3!)^2}\frac{x^3}{2^3}$$

$$\left.+K_m(n^2-\tfrac{1}{4})(n^2-\tfrac{9}{4})\dots(n^2-\tfrac{1}{4}(2m-1)^2)\frac{1}{(m!)^2}\frac{x^m}{2^m}+\dots\right\} \quad (25).$$

where $$K_m = 2\left(1+\tfrac{1}{2}+\dots\frac{1}{m}\right) \text{ so that}$$

$$K_1 = 2,\ K_2 = 3,\ K_3 = \tfrac{11}{3},\ K_4 = \tfrac{25}{6},\ K_5 = \tfrac{137}{30} \text{ \&c.}$$

If $\mathfrak{S}_n$ denote the sum of the reciprocals of the natural numbers up to and including n, then

$$1+\tfrac{1}{3}+\tfrac{1}{5}+\ldots\frac{1}{2n-1}=\mathfrak{S}_{2n}-\tfrac{1}{2}\mathfrak{S}_n=f(n)\text{ say.}$$

Now $$\mathfrak{S}_n=.577215+\log n+\frac{1}{2n}-\frac{1}{12n^2}+\ldots,$$

see Boole's *Finite Differences*, 2nd edition, p. 93.

Thus $$f(n)=.288607+\log 2n-\tfrac{1}{2}\log n+\frac{1}{48n^2}+\ldots.$$

We only want the value of b_n when x is very small, and thus we have approximately

$$b_n=\frac{\sqrt{2}}{\pi}\{1-\tfrac{1}{2}(\tfrac{1}{4}-n^2)x\}\left\{\log\frac{16\,(2+x)}{x}-4f(n)\right\}$$
$$-\frac{\sqrt{2}}{\pi}x\,(\tfrac{1}{4}-n^2)\ldots(26).$$

By equation (22)

$$c_n=\frac{2n+1}{(q^2-1)}(qb_n-b_{n+1}).$$

If we substitute for b_n and b_{n+1} their values, as given above, we find that approximately

$$c_n=\frac{\sqrt{2}}{\pi}\left[\frac{2}{x}-(n^2-\tfrac{1}{4})\left\{\log\frac{(2+x)\,16}{x}-4f(n)\right\}-(n^2+\tfrac{3}{4})\right]\ldots(27).$$

The integrals we have to evaluate are of the form

$$\frac{1}{\pi}\int_0^{2\pi}\frac{\cos n\theta\,.\,d\theta}{(R^2+\rho^2+\zeta^2-2R\rho\cos\theta)^{\frac{3}{2}}},$$

which may be written

$$\frac{1}{\pi\,(2R\rho)^{\frac{3}{2}}}\int_0^{2\pi}\frac{\cos n\theta\,.\,d\theta}{(q-\cos\theta)^{\frac{3}{2}}},$$

where $$q=\frac{R^2+\rho^2+\zeta^2}{2R\rho};$$

therefore $$x=q-1=\frac{\{(R-\rho)^2+\zeta^2\}}{2R\rho},$$

and $$2+x=1+q=\frac{(R+\rho)^2+\zeta^2}{2R\rho}.$$

Thus $$\frac{2+x}{x}=\frac{(R+\rho)^2+\zeta^2}{(R-\rho)^2+\zeta^2};$$

and the integral we wish to find $= \frac{1}{(2R\rho)^{\frac{1}{2}}} c_n$, if the value

$$x = \frac{\{(R-\rho)^2 + \zeta^2\}}{2R\rho}$$

be put for x in equation (27).

Let us denote $\frac{1}{(2R\rho)^{\frac{1}{2}}} c_n$, when x has this value by S'_n.

Then

$$\frac{1}{\{R^2 + \rho^2 + \zeta^2 - 2R\rho\cos(\psi - \phi)\}^{\frac{1}{2}}}$$
$$= S'_0 + S'_1 \cos(\psi - \phi) + \ldots S'_n \cos n(\psi - \phi) + \ldots.$$

Now in S'_n, ρ and ζ are functions of ψ,

$$\rho = a + \alpha_n \cos n\psi + \beta_n \sin n\psi,$$

and $$\zeta = z - z' = (z - \mathfrak{z}) - (\gamma_n \cos n\psi + \delta_n \sin n\psi).$$

Now let S_n be the value of S'_n when $\rho = a$ and $\zeta = (z - \mathfrak{z})$.

By Taylor's theorem,

$$S_n = S_n + (\alpha_n \cos n\psi + \beta_n \sin n\psi)\frac{dS_n}{da} - (\gamma_n \cos n\psi + \delta_n \sin n\psi)\frac{dS_n}{dz}$$
$$+ \tfrac{1}{2}(\alpha_n \cos n\psi + \beta_n \sin n\psi)^2 \frac{d^2 S_n}{da^2}$$
$$- (\alpha_n \cos n\psi + \beta_n \sin n\psi)(\gamma_n \cos n\psi + \delta_n \sin n\psi)\frac{d^2 S_n}{da\,dz}$$
$$+ \tfrac{1}{2}(\gamma_n \cos n\psi + \delta_n \sin n\psi)^2 \frac{d^2 S}{dz^2}$$

+ terms involving the cubes and higher powers of α_n, &c.

$$S'_n = S_n + \tfrac{1}{4}(\alpha_n^2 + \beta_n^2)\frac{d^2 S_n}{da^2} - \tfrac{1}{2}(\alpha_n\gamma_n + \beta_n\delta_n)\frac{d^2 S_n}{dadz}$$
$$+ \tfrac{1}{4}(\gamma_n^2 + \delta_n^2)\frac{d^2 S_n}{dz^2}$$
$$+ \cos n\psi \left\{ \alpha_n \frac{dS_n}{da} - \gamma_n \frac{dS}{dz} \right\} + \sin n\psi \left\{ \beta_n \frac{dS_n}{da} - \delta_n \frac{dS_n}{dz} \right\}$$
$$+ \tfrac{1}{4}\cos 2n\psi \left\{ (\alpha_n^2 - \beta_n^2)\frac{d^2 S_n}{da^2} - 2(\alpha_n\gamma_n - \beta_n\delta_n)\frac{d^2 S_n}{dadz} \right.$$
$$\left. + (\gamma_n^2 - \delta_n^2)\frac{d^2 S_n}{dz^2} \right\}$$
$$+ \tfrac{1}{2}\sin 2n\psi \left\{ \alpha_n\beta_n \frac{d^2 S_n}{da^2} - (\alpha_n\delta_n + \beta_n\gamma_n)\frac{d^2 S_n}{dadz} + \gamma_n\delta_n \frac{d^2 S_n}{dz^2} \right\}$$

+ terms of higher orders.

Hence, comparing these equations with § 8, we see that

$$\left.\begin{aligned}
A_n &= S_n + \tfrac{1}{4}(\alpha_n^2+\beta_n^2)\frac{d^2S_n}{da^2} - \tfrac{1}{2}(\alpha_n\gamma_n+\beta_n\delta_n)\frac{d^2S_n}{dadz} + \tfrac{1}{4}(\gamma_n^2+\delta_n^2)\frac{d^2S_n}{dz^2}\\
B_n &= \alpha_n\frac{dS_n}{da} - \gamma_n\frac{dS_n}{dz}\\
C_n &= \beta_n\frac{dS_n}{da} - \delta_n\frac{dS_n}{dz}\\
D_n &= \tfrac{1}{4}\left\{(\alpha_n^2-\beta_n^2)\frac{d^2S_n}{da^2} - 2(\alpha_n\gamma_n-\beta_n\delta_n)\frac{d^2S_n}{dadz} + (\gamma_n^2-\delta_n^2)\frac{d^2S_n}{dz^2}\right\}\\
E_n &= \tfrac{1}{2}\left\{\alpha_n\beta_n\frac{d^2S_n}{da^2} - (\alpha_n\delta_n+\beta_n\gamma_n)\frac{d^2S_n}{dadz} + \gamma_n\delta_n\frac{d^2S_n}{dz^2}\right\}
\end{aligned}\right\} \quad (28).$$

§ 13. We can now go on to find the motion of a vortex ring disturbed slightly from its circular form. It will here be only necessary to retain the first power of the quantities $\alpha_n\ldots$, so that we shall neglect all terms containing the squares of those quantities.

Fig. 2.

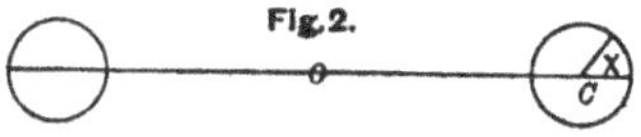

Let fig. 2 represent a section of the vortex ring by the plane of the paper. Let O be the origin of coordinates, and let C be the centre of the transverse section of the vortex core; let the radius CP of this section $= e$; let CP make an angle χ with OC produced.

Then the equations to the surface of the vortex ring are

$$\rho = a + \alpha_n\cos n\psi + \beta_n\sin n\psi + e\cos\chi \ldots\ldots\ldots (29),$$

$$z = \mathfrak{z} + \gamma_n\cos n\psi + \delta_n\sin n\psi + e\sin\chi \ldots\ldots\ldots (30).$$

Now if $F(x, y, z, t) = 0$ be an equation to a surface which as it moves always consists of the same particles of the fluid, then we know that

$$\frac{dF}{dt} + u\frac{dF}{dx} + v\frac{dF}{dy} + w\frac{dF}{dz} = 0,$$

where the differential coefficients are partial, and where u, v, w are the x, y, z components of the velocity of the fluid at the point (x, y, z).

The surface of a vortex ring is evidently a surface of this kind; we may therefore apply this result to its equation.

If we apply this theorem to equation (29), we find

$$\frac{d\alpha_n}{dt}\cos n\psi + \frac{d\beta_n}{dt}\sin n\psi - n(\gamma_n\sin n\psi - \beta_n\cos n\psi)\Psi - e\sin\chi.\mathrm{X} - \mathfrak{R} = 0,$$

where $\mathfrak{R}$ is the velocity of the fluid along the radius vector, Ψ the angular velocity of the fluid round the normals to the vortex ring, X the angular velocity round a tangent to the central line of the vortex core.

Now if the vortex be truly circular, Ψ vanishes; thus Ψ contains α_n and β_n to the first power; and $\alpha_n\Psi$ will be of the second order in α_n, and may for our present purpose be neglected. Neglecting such terms, the equation becomes

$$\frac{d\alpha_n}{dt}\cos n\psi + \frac{d\beta_n}{dt}\sin n\psi - e\sin\chi \,.\, \mathrm{X} = \mathfrak{R} \quad\ldots\ldots\ldots(31).$$

But $$u\cos\psi + v\sin\psi = \mathfrak{R}.$$

Since R and ζ are now the coordinates of a point on the surface of the vortex ring,

we have
$$R = a + \alpha_n\cos n\psi + \beta_n\sin n\psi + e\cos\chi,$$
$$\zeta = \gamma_n\cos n\psi + \delta_n\sin n\psi + e\sin\chi,$$

and writing ψ instead of ϕ in equations (11) and (14), we find, neglecting terms of the order α_n^2,

$$\begin{aligned} u\cos\psi + v\sin\psi &= \tfrac{1}{2}ma\,(\gamma_n\cos n\psi + \delta_n\sin n\psi + e\sin\chi)\,A_1 \\ &\quad + \tfrac{1}{4}m\,\{(n-1)\,A_{n+1} - (n+1)\,A_{n-1}\}\,a\gamma_n\cos n\psi \\ &\quad + \tfrac{1}{4}m\,\{(n-1)\,A_{n+1} - (n+1)\,A_{n-1}\}\,a\delta_n\sin n\psi \\ &= \tfrac{1}{2}meA_1\sin\chi + \tfrac{1}{4}\,ma\,\{2A_1 + (n-1)\,A_{n+1} \\ &\qquad\qquad - (n+1)\,A_{n+1}\}\,(\gamma_n\cos n\psi + \delta_n\sin n\psi). \end{aligned}$$

But $$\mathfrak{R} = \frac{d\alpha_n}{dt}\cos n\psi + \frac{d\beta_n}{dt}\sin n\psi - e\sin\chi\,.\,\mathrm{X}\,;$$

therefore, equating coefficients of $\sin\chi$, $\cos n\psi$, $\sin n\psi$, we get

$$\mathrm{X} = -\tfrac{1}{2}mA_1 \quad\ldots\ldots\ldots(32),$$

$$\frac{d\alpha_n}{dt} = \tfrac{1}{2}ma\gamma_n\,[A_1 + \tfrac{1}{2}\,\{(n-1)\,A_{n+1} - (n+1)\,A_{n-1}\}]\ldots(33),$$

$$\frac{d\beta_n}{dt} = \tfrac{1}{2}ma\delta_n\,[A_1 + \tfrac{1}{2}\,\{(n-1)\,A_{n+1} - (n+1)\,A_{n-1}\}]\ldots(34).$$

Now as we neglect the squares of $\alpha_n\ldots$, we may put $A_n = S_n$ and $R = a + e\cos\chi$, $\zeta = e\sin\chi$; that is, $x = \dfrac{e^2}{2a^2}$ in the quantity denoted by S_n.

Making these substitutions in equation (27), we get

$$S_n = \frac{1}{2\pi a^2}\left[\frac{4a^2}{e^2} - (n^2 - \tfrac{1}{4})\left\{\log\frac{64a^2}{e^2} - 4f(n)\right\} - (n^2 + \tfrac{3}{4})\right]\ldots(35)\,;$$

thus $$S_1 = \frac{1}{2\pi a^3}\left\{\frac{4a^2}{e^2} - \tfrac{3}{4}\left(\log\frac{64a^2}{e^2} - \tfrac{5}{3}\right)\right\};$$

therefore $$X = -\frac{m}{\pi e^2} + \frac{3m}{16\pi a^2}\left(\log\frac{64a^2}{e^2} - \tfrac{5}{3}\right);$$

or, if ω be the angular velocity of molecular rotation, so that $m = \omega\pi e^2$,

$$X = -\omega + \tfrac{3}{16}\,\omega\,\frac{e^2}{a^2}\left(\log\frac{64a^2}{e^2} - \tfrac{5}{3}\right) \quad\ldots\ldots\ldots\ldots(36),$$

and since $\frac{e}{a}$ is small, $\frac{e^2}{a^2}\log\frac{64a^2}{e^2}$ will be small; thus we have approximately

$$X = -\omega,$$

which agrees with the result given by Sir William Thomson in a note to Professor Tait's translation of Helmholtz's paper, *Phil. Mag.* 1867.

Substituting in equation (33) the values of A_1, A_{n+1}, A_{n-1}, *i.e.* in this case S_1, S_{n-1}, S_{n+1} given in equation (35), we find

$$\frac{d\alpha_n}{dt} = -\tfrac{1}{4}\,\frac{m\gamma_n}{\pi a^2}\,n^2\left\{\log\frac{64a^2}{e^2} - 4f(n) - 1\right\} \quad\ldots\ldots\ldots(37),$$

where we have neglected terms of the form $Af(n) + C$, where A and C are numerical coefficients, since when n is small $f(n)$ is small compared with $n^2\log\frac{64a^2}{e^2}$, and when n is large it is small compared with $n^2 f(n)$.

Now unless n be very large, $\log\frac{64a^2}{e^2}$ is very large compared with $f(n)$, and the equation becomes

$$\frac{d\alpha_n}{dt} = -\tfrac{1}{4}\,\frac{m\gamma_n}{\pi a^2}\,n^2\log\frac{64a^2}{e^2} \quad\ldots\ldots\ldots\ldots\ldots\ldots(38).$$

But if $f(n)$ be so large that $f(n)$ is comparable with $\log\frac{64a^2}{e^2}$; then, since approximately

$f(n) = \cdot 288607 + \log 2n - \frac{1}{2}\log n$ (Boole's *Finite Differences*, p. 93)

equation (37) becomes

$$\frac{d\alpha_n}{dt} = -\tfrac{1}{4}\,\frac{m\gamma_n}{\pi a^2}\,n^2\left(\log\frac{4a^2}{n^2e^2} - 2\cdot 1544\right) \quad\ldots\ldots\ldots\ldots(39).$$

This formula must be used when n is so large that ne is comparable with a.

We have exactly the same relation between $d\beta_n/dt$ and δ_n as between $d\alpha_n/dt$ and γ_n.

If we make the second of the equations to the surface of the vortex ring satisfy the condition necessary for it to be the equation to a surface which always consists of the same particles, we get, using the same notation as before,

$$\frac{d\mathfrak{z}}{dt}+\frac{d\gamma_n}{dt}\cos n\psi+\frac{d\delta_n}{dt}\sin n\psi-n(\gamma_n\sin n\psi-\delta_n\cos n\psi)\Psi+e\cos\chi.\mathrm{X}-w=0;$$

or, neglecting $(\gamma_n \sin n\psi - \delta_n \cos n\psi)\Psi$ as before we find

$$\frac{d\mathfrak{z}}{dt}+\frac{d\gamma_n}{dt}\cos n\psi+\frac{d\delta_n}{dt}\sin n\psi+e\cos\chi.\mathrm{X}=w \quad\ldots\ldots(40).$$

But we know by equations (16) and (17) that

$$\begin{aligned}w=\tfrac{1}{2}m(2a^2A_0-aRA_1)&\\+\tfrac{1}{2}m[a^2B_n-\tfrac{1}{2}aR(B_{n+1}+B_{n+1})+\tfrac{1}{2}\{(n-1)A_{n-1}-(n+1)A_{n+1}\}R\alpha_n&\\+2aA_n\alpha_n]\cos n\psi&\\+\tfrac{1}{2}m[a^2C_n-\tfrac{1}{2}aR(C_{n+1}+C_{n-1})+\tfrac{1}{2}\{(n-1)A_{n-1}-(n+1)A_{n+1}\}R\beta_n&\\+2aA_n\beta_n]\sin n\psi,&\end{aligned}$$

where $R=a+\alpha_n\cos n\psi+\beta_n\sin n\psi+e\cos\chi$, &c.

Substituting this value for R and the values of A_n, B_n, &c. given in equation (28), we find

$$\begin{aligned}w=\tfrac{1}{2}m(2a^2S_0-a^2S_1)-\tfrac{1}{2}mae\cos\chi.S_1&\\+\tfrac{1}{2}m\Big[a\frac{d}{da}\{S_n-\tfrac{1}{2}(S_{n+1}+S_{n-1})\}&\\+\tfrac{1}{2}\{(n-1)S_{n-1}-(n+1)S_{n+1}\}+2S_n&\\-S_1+a\frac{d}{dR}(2S_0-S_1)\Big]\,a\,(\alpha_n\cos n\psi+\beta_n\sin n\psi).&\end{aligned}$$

Where in S_n, R after differentiation is put equal to $a+e\cos\chi$, and $x=\dfrac{e^2}{2a^2}$,

Equating in the two expressions for w, the term independent of ψ and χ, the coefficient of $\cos\chi$ and the coefficients of $\cos n\psi$ and $\sin n\psi$, we get

$$\frac{d\mathfrak{z}}{dt} = \tfrac{1}{2}m\,(2a^2S_0 - a^2S_1),$$

$$X = -\tfrac{1}{2}m\,ae\,S_1,$$

$$\frac{d\gamma_n}{dt} = \tfrac{1}{2}m\,a\alpha_n\left[a\frac{d}{da}\{S_n - \tfrac{1}{2}(S_{n+1} + S_{n-1})\}\right.$$
$$+\tfrac{1}{2}\{(n-1)S_{n-1} - (n+1)S_{n+1}\} + 2S_n - S_1$$
$$\left.+\,a\frac{d}{dR}(2S_0 - S_1)\right],$$

with a similar equation between $d\delta_n/dt$ and β_n.

S_n before differentiation

$$= \frac{1}{2\pi(Ra)^{\frac{3}{2}}}\left[\frac{2}{x} - (n^2 - \tfrac{1}{4})\left\{\log\left(\frac{2+x}{x}16\right) - 4f(n)\right\} - (n^2 + \tfrac{3}{4})\right],$$

where $$x = \frac{(R-a)^2 + \zeta^2}{2Ra}.$$

When S_n has not to be differentiated, it equals

$$\frac{1}{2\pi a^3}\left[\frac{4a^2}{e^2} - (n^2 - \tfrac{1}{4})\left\{\log\frac{64a^2}{e^2} - 4f(n)\right\} - (n^2 + \tfrac{3}{4})\right].$$

The first equation gives the velocity of translation of the vortex ring, substituting the values for S_0 and S_1 we find

$$\frac{d\mathfrak{z}}{dt} = \frac{m}{4\pi a}\left(\log\frac{64a^2}{e^2} - 2\right)$$
$$= \frac{m}{2\pi a}\left(\log\frac{8a}{e} - 1\right)\text{.....................}(41).$$

In a note to Professor Tait's translation of Helmholtz's paper on Vortex Motion, *Phil. Mag.*, 1867, Sir William Thomson states that the velocity of translation of a circular vortex ring is

$$\frac{m}{2\pi a}\left(\log\frac{8a}{e} - \tfrac{1}{4}\right).$$

This agrees very approximately with the result we have just obtained, and Mr T. C. Lewis, in the *Quarterly Journal of Mathematics*, vol. XVI. obtains the same expression as we have for the velocity of translation.

The second expression gives the same value for the angular velocity X as we had before.

The third equation gives on substitution and differentiation

$$\frac{d\gamma_n}{dt} = \frac{m\alpha_n}{4\pi a^3}(n^2 - 1)\left\{\log\frac{64a^2}{e^2} - 4f(n) - 1\right\}\text{..........}(42),$$

neglecting as before terms of the form $Af(n) + C$, where A and C are numerical coefficients.

We have a similar equation between $d\delta_n/dt$ and β_n.

Substituting these values for $\frac{d\gamma_n}{dt}$ and $\frac{d\delta_n}{dt}$ in equation (40), we find that the velocity of translation W at any point on the ring is given by

$$W = \frac{d\mathfrak{z}}{dt} + \frac{m}{4\pi a^2}(n^2-1)\left\{\log\frac{64a^2}{e^2} - 4f(n) - 1\right\}(\alpha_n \cos n\psi + \beta_n \sin n\psi);$$

or, neglecting $4f(n)$,

$$W = \frac{d\mathfrak{z}}{dt}\left\{1 + \frac{n^2-1}{a}(\alpha_n \cos n\psi + \beta_n \sin n\psi)\right\}.$$

If ρ' be the radius of curvature at any point of the central line of vortex core, we can easily prove that

$$\frac{1}{\rho'} = \frac{1}{a} + \frac{n^2-1}{a^2}(\alpha_n \cos n\psi + \beta_n \sin n\psi),$$

so that the velocity of translation of any point of the vortex ring

$$= \frac{d\mathfrak{z}}{dt}\,\frac{a}{\rho'};$$

thus those portions of the axis which at any time have the greatest curvature will have the greatest velocity.

Returning to the equation for $\frac{d\gamma_n}{dt}$, we have as before

$$\frac{d\gamma_n}{dt} = (n^2-1)L\alpha_n \quad\text{.......................(43)},$$

where
$$L = \frac{m}{4\pi a^2}\log\frac{64a^2}{e^2},$$

except when n is so large that ne is at all comparable with a, then

$$L = \frac{m}{4\pi a^2}\left(\log\frac{4a^2}{n^2e^2} - 2{\cdot}1544\right),$$

approximately; the accurate value of L is

$$\log\frac{64a^2}{e^2} - 4f(n) - 1;$$

this is the same coefficient as we had in the equation giving $d\alpha_n/dt$ so that our equations are

$$\frac{d\alpha_n}{dt} = -n^2L\gamma_n,$$

$$\frac{d\gamma_n}{dt} = (n^2-1)L\alpha_n.$$

Differentiating the first of these, and substituting for $\frac{d\gamma_n}{dt}$ from the second, we find

$$\frac{d^2\alpha_n}{dt^2} + n^2(n^2-1)L^2\alpha_n = 0,$$

the solution of which is

$$\left.\begin{aligned} &\alpha_n = A\cos[L\sqrt{\{n^2(n^2-1)\}}\,t + B] \\ &\text{and therefore} \\ &\gamma_n = A\sqrt{\left(\frac{n^2-1}{n^2}\right)}\sin[L\sqrt{\{n^2(n^2-1)\}}\,t + B] \end{aligned}\right\}\ldots\ldots\ldots\ldots(44),$$

where A and B are arbitrary constants.

We can shew by work of an exactly similar kind, that

$$\left.\begin{aligned} \beta_n &= A'\cos[L\sqrt{\{n^2(n^2-1)\}}\,t + B'] \\ \delta_n &= A'\sqrt{\left(\frac{n^2-1}{n^2}\right)}\sin[L\sqrt{\{n^2(n^2-1)\}}\,t + B'] \end{aligned}\right\}\ldots\ldots\ldots(45).$$

These equations shew that the circular vortex ring is stable for all small displacements of its central line of vortex core. Sir William Thomson has proved, that it is stable for all small alterations in the shape of its transverse section, hence we conclude that it is stable for all small displacements. The time of vibration

$$= \frac{2\pi}{L\sqrt{\{n^2(n^2-1)\}}},$$

$$= \frac{2\pi}{\sqrt{\{n^2(n^2-1)\}}} \cdot \frac{4\pi a^2}{m\left(\log\frac{64a^2}{e^2} - 4f(n) - 1\right)}\ldots\ldots\ldots(46),$$

where $$f(n) = 1 + \tfrac{1}{3} + \tfrac{1}{5} + \ldots\frac{1}{2n-1}.$$

Thus, unless n be very large, the time of vibration

$$= \frac{2\pi}{\sqrt{\{n^2(n^2-1)\}}}\,\frac{2\pi a^2}{m\log\frac{8a}{e}},$$

or, if V be the velocity of translation of the vortex ring

$$= \frac{2\pi}{\sqrt{\{n^2(n^2-1)\}}}\,\frac{a}{V}.$$

Thus for elliptic deformation the time of vibration is ·289 times the time taken by the vortex ring to pass over a length equal to its circumference.

When ne is at all comparable with a, the time of vibration is approximately

$$\frac{2\pi}{\sqrt{\{n^2(n^2-1)\}}}\cdot\frac{2\pi a^2}{m\left(\log\frac{2a}{ne}-1\cdot0772\right)};$$

or, since we may write, as n is large, n^2 instead of n^2-1, it equals, if l be the wave length $\frac{2\pi a}{n}$,

$$\frac{2\pi}{\frac{2\omega\pi^3e^2}{l^2}\left(\log\frac{l}{\pi e}-1\cdot0772\right)}.$$

Now this case agrees infinitely nearly with the transverse vibrations of a straight columnar vortex which have been investigated by Sir William Thomson.

In the sub-case in which l/e is large, he finds that the period of vibration

$$=\frac{2\pi}{\frac{2\omega\pi^3e^2}{l^2}\left(\log\frac{l}{2\pi e}+\cdot1159+\frac{1}{4}\right)}$$

(*Phil. Mag.*, Sept. 1880, p. 167 eq. 61); or, since $\log_e 2=\cdot62314$, this equals

$$\frac{2\pi}{\frac{2\omega\pi^3e^2}{l^2}\left(\log\frac{l}{\pi e}-\cdot3272\right)},$$

and thus agrees very approximately with the value we have just found.

Since the amplitudes of α_n and β_n when n is large are approximately the same as those of γ_n and δ_n, we can represent a displacement of this kind by conceiving the central line of the vortex core to be wound round an anchor ring of small transverse section so as to make n turns round the central line of the vortex ring, and this form to travel along the anchor ring with velocity $\frac{l}{\tau}$, where τ is the time of vibration just found and l the wave length.

PART II.

To find the action of two vortices upon each other which move so as never to approach closer than a large multiple of the diameter of either.

§ 14. The expressions for the velocity due to a circular vortex ring, which we investigated in the previous part, will enable us to solve this problem. If we call the two vortices AB and CD, then in order to find the effect of the vortex AB on CD we must find the velocity at CD due to AB. Now, since the vortices never approach very closely to each other, they will not differ much from circles; hence in finding the velocity due to one of them at a point remote from its core, say at the surface of the other, we may, without appreciable error, suppose the vortex ring to be circular.

Let the shortest distance between the directions of motion of the vortices be perpendicular to the plane of the paper; thus the plane of the paper will be parallel to the directions of motion of both vortices.

Let the semi-polar equations to the central line of the vortex AB of strength m (fig. 3) be

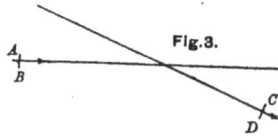

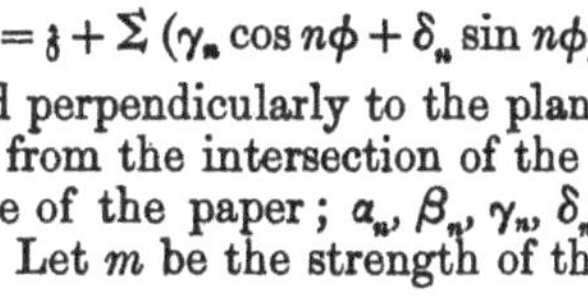

Fig. 3.

$$\rho = a + \Sigma (\alpha_n \cos n\phi + \beta_n \sin n\phi),$$
$$z = \mathfrak{z} + \Sigma (\gamma_n \cos n\phi + \delta_n \sin n\phi),$$

when z is measured perpendicularly to the plane of the vortex AB and ϕ is measured from the intersection of the plane of the vortex AB with the plane of the paper; α_n, β_n, γ_n, δ_n are all very small compared with a. Let m be the strength of the vortex AB.

Let the equations to the central line of the vortex CD of strength m' be

$$\rho' = b + \Sigma (\alpha'_n \cos n\psi' + \beta'_n \sin n\psi'),$$
$$z' = \mathfrak{z}' + \Sigma (\gamma'_n \cos n\psi' + \delta'_n \sin n\psi'),$$

where z' is measured perpendicularly to the plane of the vortex CD, and ψ' from the intersection of the plane of this vortex with the plane of the paper; α'_n, β'_n, γ'_n, δ'_n are all very small in comparison with b.

We shall have to express α_n, β_n, γ_n, δ_n, α'_n, β'_n, γ'_n, δ'_n as functions of the time; we shall then have found the action of the two vortices on each other.

To find the action of AB on CD let us take as our axis of Z the perpendicular to the plane of the vortex AB through its centre, the plane of XZ parallel to the plane of the paper and the axis of Y drawn upwards from the plane of the paper.

Let ϵ be the angle between the direction of motion of the two vortices; l, m, n the direction-cosines of a radius vector of the vortex CD drawn from the centre of that vortex.

Let Z, X (fig. 4) be the points where the axes of Z and X cut

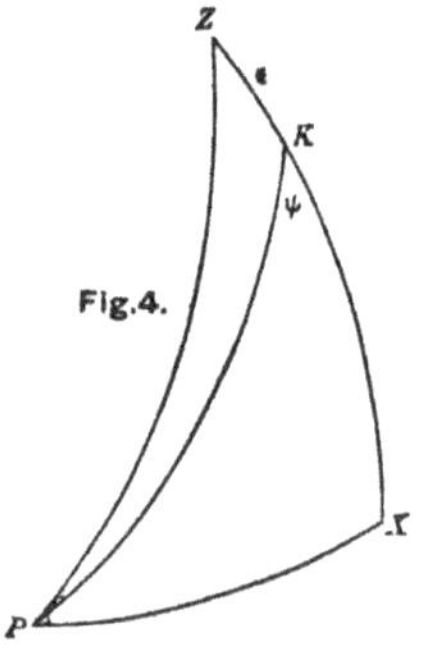

Fig. 4.

a sphere whose centre is at the origin of coordinates, K the point where a parallel to the direction of motion of the vortex CD cuts this sphere, and P the point where a parallel to the radius vector of the vortex CD cuts the sphere: KP will be a quadrant of a circle.

Then we easily see, by Spherical Trigonometry, that

$$l = \cos\epsilon\cos\psi,$$

$$m = \sin\psi,$$

$$n = -\sin\epsilon\cos\psi.$$

Now by equations (10, 13, 16) the velocities u, v, w parallel to the axes of X, Y, Z due to the vortex AB supposed circular are given by the equations

$$u = \frac{1}{2R} maZXA_1,$$
$$v = \frac{1}{2R} maZYA_1,$$
$$w = \tfrac{1}{2}m(2a^2A_0 - aRA_1),$$
where
$$R = \sqrt{X^2 + Y^2}.$$

Since

$$\frac{1}{(a^2+R^2+Z^2-2aR\cos\theta)^{\frac{3}{2}}} = \frac{1}{(R^2+Z^2)^{\frac{3}{2}}} - \tfrac{3}{2}\frac{(a^2-2aR\cos\theta)}{(R^2+Z^2)^{\frac{5}{2}}} + \tfrac{15}{8}\frac{4a^2R^2\cos^2\theta}{(R^2+Z^2)^{\frac{7}{2}}} + \dots,$$

where, since R^2+Z^2 is very great compared with a, the terms diminish rapidly,

$$A_0 = \frac{1}{(R^2+Z^2)^{\frac{3}{2}}} - \tfrac{3}{2}\frac{a^2}{(R^2+Z^2)^{\frac{5}{2}}} + \tfrac{15}{4}\frac{a^2R^2}{(R^2+Z^2)^{\frac{7}{2}}} + \dots,$$

and
$$A_1 = \frac{3aR}{(R^2+Z^2)^{\frac{5}{2}}}.$$

Now if f, g, h be the coordinates of the centre of the vortex CD, and X, Y, Z the coordinates of a point on the central line of that vortex,

$$X = f + bl = f + b\cos\epsilon\cos\psi,$$
$$Y = g + bm = g + b\sin\psi,$$
$$Z = h + bn = h - b\sin\epsilon\cos\psi;$$

therefore

$$R^2 + Z^2 = X^2 + Y^2 + Z^2$$
$$= f^2 + g^2 + h^2 + 2b(f\cos\epsilon\cos\psi + g\sin\psi - h\sin\epsilon\cos\psi) + b^2.$$

§ 15. $$u = \frac{1}{2R}maXZA_1 = \tfrac{3}{2}ma^2\frac{XZ}{(X^2+Y^2+Z^2)^{\frac{5}{2}}}.$$

Substituting the values given above for X, Y, Z and writing d^2 for $f^2+g^2+h^2+b^2$ we find that approximately

$$u = \tfrac{3}{2}ma^2$$
$$\times\left[\frac{fh}{d^5} + \left(h\cos\epsilon - f\sin\epsilon - \frac{5fh}{d^2}(f\cos\epsilon - h\sin\epsilon)\right)\frac{b\cos\psi}{d^5} - \frac{5fgh\sin\psi}{d^7}\right.$$
$$+\left(-\sin\epsilon\cos\epsilon - \frac{5}{d^2}(h\cos\epsilon - f\sin\epsilon)(f\cos\epsilon - h\sin\epsilon)\right.$$
$$\left.+\frac{35}{2d^4}fh\{(f\cos\epsilon - h\sin\epsilon)^2 - g^2\}\right)\frac{b^2}{2d^5}\cos 2\psi$$
$$\left.+\left(f\sin\epsilon - h\cos\epsilon + \frac{7}{d^2}fg(f\cos\epsilon - h\sin\epsilon)\right)\frac{5b^2}{2d^7}\sin 2\psi + \dots\right] \dots (47).$$

When in these expressions we have a coefficient consisting of several terms of different orders of small quantities we only retain the largest term.

§ 16. $$v = \frac{1}{2R} maYZA_1 = \tfrac{3}{2} ma^2 \frac{YZ}{(X^2+Y^2+Z^2)^{\frac{5}{2}}}.$$

Substituting as before we find

$$v = \tfrac{3}{2} ma^2$$

$$\times \left[\frac{gh}{d^5} - \left(g \sin\epsilon + \frac{5hg}{d^2}(f\cos\epsilon - h\sin\epsilon)\right)\frac{b}{d^5}\cos\psi \right.$$

$$+ h\left(1 - \frac{5g^2}{d^2}\right)\frac{b}{d^5}\sin\psi$$

$$+ \tfrac{5}{2}\left(\cos\epsilon\, g\,(f\sin\epsilon - h\cos\epsilon) + \tfrac{7}{2}\frac{gh}{d^2}\{(f\cos\epsilon - h\sin\epsilon)^2 - g^2\}\right)\frac{b^2}{d^7}\cos 2\psi$$

$$+ \left(\left(\frac{5g^2}{d^2} - 1\right)\sin\epsilon - \frac{5h}{d^2}(f\cos\epsilon - h\sin\epsilon)\right.$$

$$\left.\left. + \frac{35hg^2}{d^4}(f\cos\epsilon - h\sin\epsilon)\right)\frac{b^2}{2d^5}\sin 2\psi\right] \text{.........(48).}$$

§ 17. $$w = \tfrac{1}{2}m\,(2a^2A_0 - aRA_1)$$

$$= \tfrac{1}{2}m\left\{\frac{2a^2}{(X^2+Y^2+Z^2)^{\frac{3}{2}}} - \frac{3a^2(a^2+R^2)}{(X^2+Y^2+Z^2)^{\frac{5}{2}}} + \tfrac{15}{4}\frac{a^4R^2}{(X^2+Y^2+Z^2)^{\frac{7}{2}}}\right\}$$

$$= \tfrac{1}{2}\frac{ma^2}{d^3} \times$$

$$\left[2 - \frac{3}{d^2}(f^2+g^2) + 3\left(2\,(h\sin\epsilon - 2f\cos\epsilon)\right.\right.$$

$$\left. + \frac{5}{d^2}(f^2+g^2)(f\cos\epsilon - h\sin\epsilon)\right)\frac{b}{d^2}\cos\psi + 3g\,(f^2+g^2-4h^2)\frac{b}{d^4}\sin\psi$$

$$+ \tfrac{3}{2}\left(\sin^2\epsilon + \frac{5}{d^2}\{(f\cos\epsilon - h\sin\epsilon)(3f\cos\epsilon - h\sin\epsilon) - g^2\}\right.$$

$$\left. - \frac{35}{2d^4}(f^2+g^2)\{(f\cos\epsilon - h\sin\epsilon)^2 - g^2\}\right)\frac{b^2}{d^2}\cos 2\psi$$

$$+ 15\left(3f\cos\epsilon - h\sin\epsilon\right.$$

$$\left.\left. - \frac{7}{2d^2}(f^2+g^2)(f\cos\epsilon - h\sin\epsilon)\right)\frac{b^2g}{d^4}\sin 2\psi + \ldots\right] \text{... (49).}$$

§ 18. In using these expressions to find the effect of the vortex AB on CD, we have to find the velocity perpendicular to the plane of CD and along the radius vector. Then, as in the case of the single vortex, we have equations of the type $\frac{da'_n}{dt}$ = coefficient of $\cos n\psi$ in the expression for the velocity along the radius vector of CD.

To solve these differential equations, we must have the quantities on the right-hand side expressed in terms of the time. Hence we must express the value for u, v, w which we have just obtained in terms of the time.

§ 19. In the small terms which express the velocity at the vortex CD due to the vortex AB, we may, for a first approximation, calculate the quantities on the supposition that the motion is undisturbed.

Let us reckon the time from the instant when the centres of the vortices are nearest together.

Let p and q be the velocities of the vortices AB and CD respectively; k the relative velocity, viz. $\sqrt{(p^2 + q^2 - 2pq \cos\epsilon)}$; c the shortest distance between their centres.

Then, since f, g, h are the coordinates of the centre of CD at the time t,

$$f = \mathfrak{f} + q \sin\epsilon \,.\, t,$$
$$g = \mathfrak{g},$$
$$h = \mathfrak{h} + (q \cos\epsilon - p)\, t,$$

where $\mathfrak{f}, \mathfrak{g}, \mathfrak{h}$ are the values of f, g, h when $t = 0$; since the distance between the centres of the vortices, viz. $\sqrt{(f^2 + g^2 + h^2)}$ is a minimum when $t = 0$,

$$\mathfrak{f} q \sin\epsilon + \mathfrak{h}\,(q \cos\epsilon - p) = 0\,;$$

therefore
$$\frac{-\mathfrak{f}}{q\cos\epsilon - p} = \frac{\mathfrak{h}}{q\sin\epsilon} = \frac{\sqrt{(c^2 - \mathfrak{g}^2)}}{k}\,;$$

therefore if $\mathfrak{h}$ be positive, we have

$$\mathfrak{h} = \frac{q \sin\epsilon \sqrt{(c^2 - \mathfrak{g}^2)}}{k} \quad\text{......................(50)},$$

$$\mathfrak{f} = -\frac{(q\cos\epsilon - p)\sqrt{(c^2 - \mathfrak{g}^2)}}{k} \quad\text{..............(51)},$$

and
$$f^2 + g^2 + h^2 = c^2 + k^2 t^2.$$

§ 20. If we substitute for f, g, h in the expression for w their values in terms of the time, we find that as far as the term independent of ψ goes,

$$w = \tfrac{1}{2}\frac{ma^2}{(c^2+k^2t^2)^{\frac{5}{2}}}\Bigg[\left(3\,(c^2-\mathfrak{g}^2)\frac{q^2\sin^2\epsilon}{k^2}-c^2\right)$$

$$+\frac{6\sqrt{c^2-\mathfrak{g}^2}}{k}(q\cos\epsilon-p)\,q\sin\epsilon . t+\{2(q\cos\epsilon-p)^2-q^2\sin^2\epsilon\}t^2\Bigg]\ldots(52).$$

The coefficient of $\cos\psi$

$$= \tfrac{3}{2}\,ma^2\left\{2\left(\frac{\sqrt{c^2-\mathfrak{g}^2}}{k}\{q\,(\sin^2\epsilon+2\cos^2\epsilon)-2p\cos\epsilon\}\right.\right.$$

$$\left.\left.-\sin\epsilon\,(q\cos\epsilon+p)\,t\right)\frac{1}{(c^2+k^2t^2)^{\frac{5}{2}}}+(L+Mt+Nt^2+Pt^3)\frac{5}{(c^2+k^2t^2)^{\frac{7}{2}}}\right\}$$

$$\ldots\ldots\ldots(53),$$

where

$$L=\frac{\sqrt{c^2-\mathfrak{g}^2}}{k^3}(p\cos\epsilon-q)\left(c^2\,(p-q\cos\epsilon)^2+\mathfrak{g}^2q^2\sin^2\epsilon\right)$$

$$M=\frac{\sin\epsilon}{k^2}\Big(c^2\,(p-q\cos\epsilon)\,(p^2+pq\cos\epsilon-2q^2)+\mathfrak{g}^2q\,(pq\,(3+\cos^2\epsilon)$$

$$-2\,(p^2+q^2)\cos\epsilon)\Big)$$

$$N=\frac{\sqrt{c^2-\mathfrak{g}^2}}{k}\sin^2\epsilon\,.\,q\,\{2p^2-qp\cos\epsilon-q^2\}$$

$$P=\sin^3\epsilon\,q^2p.$$

The coefficient of $\sin\psi$

$$=\frac{3ma^2b\mathfrak{g}}{2\,(c^2+k^2t^2)^{\frac{7}{2}}}\{c^2-5\frac{(c^2-\mathfrak{g}^2)}{k^2}q^2\sin^2\epsilon$$

$$+\frac{10\sqrt{c^2-\mathfrak{g}^2}}{k}q\sin\epsilon\,(p-q\cos\epsilon)\,t+(5q^2\sin^2\epsilon-4k^2)t^2\}\ldots\ldots(54).$$

The coefficient of $\cos\psi$ may be written

$$\tfrac{3}{2}\,ma^2b^2\left\{\frac{\sin^2\epsilon}{(c^2+k^2t^2)^{\frac{5}{2}}}+\left(\frac{c^2-\mathfrak{g}^2}{k^2}(p\cos\epsilon-q)\,(3\cos\epsilon\,(p-q\cos\epsilon)-q\sin^2\epsilon)-\mathfrak{g}^2\right.\right.$$

$$+\frac{2\sqrt{c^2-\mathfrak{g}^2}}{k}\sin\epsilon\,\{2p\,(p\cos\epsilon-q)+q\,(p-q\cos\epsilon)\}\,t$$

$$\left.+\sin\epsilon\,p\,(q\sin 2\epsilon+p\sin\epsilon)\,t^2\right)\frac{1}{(c^2+k^2t^2)^{\frac{7}{2}}}$$

$$\left.-\tfrac{35}{4}\,(L'+M't+N't^2+P't^3+Q't^4)\frac{1}{(c^2+k^2t^2)^{\frac{9}{2}}}\right\}\ldots\ldots\ldots(55),$$

where

$$L' = \frac{1}{k^4}(c^2(q\cos\epsilon - p)^2 + \mathfrak{g}^2 q^2 \sin^2\epsilon)(c^2(p\cos\epsilon - q)^2 - \mathfrak{g}^2\{(p\cos\epsilon - q)^2 + k^2\}),$$

$$M' = \frac{2\sqrt{c^2-\mathfrak{g}^2}}{k^3}\sin\epsilon\{c^2(p^2-q^2)(p\cos\epsilon - q)(p - q\cos\epsilon) + \mathfrak{g}^2 q((p\cos\epsilon - q)(2pq - \cos\epsilon(p^2+q^2)) + (q\cos\epsilon - p)(p^2+q^2-2pq\cos\epsilon))\},$$

$$N' = \frac{\sin^2\epsilon}{k^2}\{c^2p^2(q\cos\epsilon - p)^2 - \mathfrak{g}^2q^2(p\cos\epsilon - q)^2 - 4pq(c^2-\mathfrak{g}^2)(p - q\cos\epsilon)(p\cos\epsilon - q)\},$$

$$P' = \frac{2\sin^3\epsilon}{k}\sqrt{c^2-\mathfrak{g}^2}\,pq(p^2-q^2),$$

$$Q' = \sin^4\epsilon\, p^2q^2.$$

The coefficient of $\sin 2\psi$ may be written

$$\tfrac{15}{2} ma^2\mathfrak{g}\left\{\left(\frac{\sqrt{c^2-\mathfrak{g}^2}}{k}\{3p\cos\epsilon - q(3\cos^2\epsilon + \sin^2\epsilon)\} + (q\sin 2\epsilon + p\sin\epsilon)t\right)\frac{1}{(c^2+k^2t^2)^{\frac{7}{2}}} - \tfrac{7}{2}(L + Mt + Nt^2 + Pt^3)\frac{1}{(c^2+k^2t^2)^{\frac{9}{2}}}\right\} \quad \text{......(56)},$$

where L, M, N, P have the same values as in equation (53).

§ 21. The velocity parallel to the axis of y.

The term independent of ψ

$$= \tfrac{3}{2}\frac{ma^2\mathfrak{g}}{(c^2+k^2t^2)^{\frac{5}{2}}}\left\{\frac{\sqrt{c^2-\mathfrak{g}^2}}{k}q\sin\epsilon + (q\cos\epsilon - p)t\right\} \quad \text{....(57)}.$$

The coefficient of $\cos\psi$

$$= -\tfrac{3}{2} ma^2b\mathfrak{g}\left\{\frac{\sin\epsilon}{(c^2+k^2t^2)^{\frac{5}{2}}} + 5\left(\frac{(c^2-\mathfrak{g}^2)}{k^2}\sin\epsilon\,.\,q(p\cos\epsilon - q) + \frac{\sqrt{c^2-\mathfrak{g}^2}}{k}\{2pq - \cos\epsilon(p^2+q^2)\}t + \sin\epsilon\,.\,p(q\cos\epsilon - p)t^2\right)\frac{1}{(c^2+k^2t^2)^{\frac{7}{2}}}\right\} \quad \text{..(58)}.$$

The coefficient of $\sin\psi$

$$= \tfrac{3}{2} ma^2b\mathfrak{g}\left(\sqrt{c^2-\mathfrak{g}^2}\frac{q\sin\epsilon}{k} + (q\cos\epsilon - p)t\right)\left(\frac{1}{(c^2+k^2t^2)^{\frac{5}{2}}} - \frac{5\mathfrak{g}^2}{(c^2+k^2t^2)^{\frac{7}{2}}}\right) \quad \text{...(59)}.$$

The coefficient of $\cos 2\psi$

$$= \tfrac{3}{2}\, ma^2b^2\mathfrak{g} \left\{ \cos\epsilon \frac{\sqrt{c^2-\mathfrak{g}^2}}{k} (p\sin\epsilon - q\sin 2\epsilon) \frac{1}{(c^2+k^2t^2)^{\frac{7}{2}}} \right.$$

$$\left. + \tfrac{7}{2} (L'' + M''t + N''t^2 + P''t^3) \frac{1}{(c^2+k^2t^2)^{\frac{9}{2}}} \right\} \ldots (60)$$

where

$$L'' = \frac{\sqrt{c^2-\mathfrak{g}^2}}{k^3} \{(c^2-\mathfrak{g}^2)(p\sin\epsilon - q)^2 - \mathfrak{g}^2 k^2\},$$

$$M'' = \frac{1}{k^2} \{(c^2-\mathfrak{g}^2)(p\cos\epsilon - q)[(2+\sin^2\epsilon)pq - \cos\epsilon(p^2+q^2)]$$

$$- \mathfrak{g}^2 k^2 (q\cos\epsilon - p)\},$$

$$N'' = \frac{\sqrt{c^2-\mathfrak{g}^2}}{k} p\sin\epsilon \{pq(3+\cos^2\epsilon) - 2(p^2+q^2)\cos\epsilon\},$$

$$P'' = \sin^2\epsilon \,.\, p^2 (q\cos\epsilon - p).$$

The coefficient of $\sin 2\psi$

$$= -\tfrac{3}{2}\, ma^2b^2 \left\{ \sin\epsilon \left(1 - \frac{5\mathfrak{g}^2}{(c^2+k^2t^2)}\right) \frac{1}{(c^2+k^2t^2)^{\frac{7}{2}}} \right.$$

$$+ 5\left(1 - \frac{7\mathfrak{g}^2}{(c^2+k^2t^2)}\right) \left(\frac{c^2-\mathfrak{g}^2}{k^2} \sin\epsilon \,.\, q(p\cos\epsilon - q) \right.$$

$$\left. \left. + \frac{\sqrt{c^2-\mathfrak{g}^2}}{k} \{2pq - \cos\epsilon(p^2+q^2)\} t + \sin\epsilon \,.\, p(q\cos\epsilon - p) t^2 \right) \frac{1}{(c^2+k^2t^2)^{\frac{7}{2}}} \right\}$$

$$\ldots\ldots\ldots\ldots\ldots\ldots\ldots\ldots (61).$$

§ 22. The velocity parallel to the axis of x.

The term independent of ψ

$$= \tfrac{3}{2}\, ma^2 \left\{ \frac{(c^2-\mathfrak{g}^2)}{k^2} \sin\epsilon \,.\, q(p - q\cos\epsilon) \right.$$

$$\left. - \frac{\sqrt{c^2-\mathfrak{g}^2}}{k} \{q^2\cos 2\epsilon - 2pq\cos\epsilon + p^2\} t + \sin\epsilon \,.\, q(q\cos\epsilon - p) t^2 \right\} \frac{1}{(c^2+k^2t^2)^{\frac{5}{2}}}$$

$$\ldots\ldots\ldots\ldots\ldots\ldots\ldots\ldots (62).$$

The coefficient of $\cos\psi$

$$= \tfrac{3}{2}\, ma^2b \left\{ \left(\frac{\sqrt{c^2-\mathfrak{g}^2}}{k} (q\sin 2\epsilon - p\sin\epsilon) + (q\cos 2\epsilon - p\cos\epsilon) t \right) \frac{1}{(c^2+k^2t^2)^{\frac{5}{2}}} \right.$$

$$\left. - 5 (L''' + M'''t + N'''t^2 + P'''t^3) \frac{1}{(c^2+k^2t^2)^{\frac{7}{2}}} \right\} \ldots (63),$$

where

$$L''' = \frac{(c^2 - \mathfrak{g}^2)^{\frac{3}{2}}}{k^3} \sin \epsilon \,.\, q\,(q \cos \epsilon - p)\,(q - p \cos \epsilon),$$

$$M''' = \frac{c^2 - \mathfrak{g}^2}{k^3}\,(q^3 \cos 2\epsilon - pq^2 \cos \epsilon\,(\cos^2 \epsilon + 2) + p^2 q\,(2 + \cos^2 \epsilon) - p^3 \cos \epsilon),$$

$$N''' = \sin \epsilon \,.\, \frac{\sqrt{c^2 - \mathfrak{g}^2}}{k}\,(-q^3 \cos \epsilon + q^2 p\,(1 + \sin^2 \epsilon) + qp^2 \cos \epsilon - p^3),$$

$$P''' = \sin^2 \epsilon \,.\, pq\,(q \cos \epsilon - p).$$

The coefficient of $\sin \psi$

$$= -\tfrac{15}{2}\, ma^2 b \mathfrak{g} \left\{ \frac{(c^2 - \mathfrak{g}^2)}{k^2} \sin \epsilon \,.\, q\,(p - q \cos \epsilon) - \frac{\sqrt{c^2 - \mathfrak{g}^2}}{k}\,(q^2 \cos 2\epsilon - 2pq \cos \epsilon + p^2)\, t + \sin \epsilon . q\,(q \cos \epsilon - p)\, t^2 \right\} \frac{1}{(c^2 + k^2 t^2)^{\frac{7}{2}}} \quad \ldots (64).$$

The coefficient of $\cos 2\psi$

$$= \tfrac{3}{8}\, ma^2 b^2 \left\{ \frac{-\sin \epsilon \cos \epsilon}{(c^2 + k^2 t^2)^{\frac{5}{2}}} - \tfrac{5}{2} \left(-\frac{(c^2 - \mathfrak{g}^2)}{k^2} \sin \epsilon\, \{2q^2 \cos \epsilon - pq\,(1 + 2 \cos^2 \epsilon) + p^2 \cos \epsilon\} - \frac{\sqrt{c^2 - \mathfrak{g}^2}}{k}(q^2 \cos 2\epsilon - 2pq \cos \epsilon + p^2) t + \sin \epsilon . p (q \cos 2\epsilon - p \cos \epsilon) t^2 \right) \frac{1}{(c^2 + k^2 t^2)^{\frac{7}{2}}} + \tfrac{35}{2}\,(L_1 + M_1 t + N_1 t^2 + P_1 t^3 + Q_1 t^4)\, \frac{1}{(c^2 + k^2 t^2)^{\frac{9}{2}}} \right\} \quad \ldots (65),$$

where

$$L_1 = -\frac{(c^2 - \mathfrak{g}^2)}{k^4} \sin \epsilon \,.\, q\,(q \cos \epsilon - p)\, \{c^2\,(p \cos \epsilon - q)^2 - \mathfrak{g}^2\,[(p \cos \epsilon - q)^2 + k^2]\},$$

$$M_1 = \frac{\sqrt{c^2 - \mathfrak{g}^2}}{k^3} \{(c^2 - \mathfrak{g}^2)\,(p \cos \epsilon - q)(q^3 \cos 2\epsilon - pq^2 \cos \epsilon\,(\cos 2\epsilon + \cos^2 \epsilon) + p^2 q\,(1 + 2 \cos 2\epsilon) - p^3 \cos \epsilon) + \mathfrak{g}^2 k^2\,(q^2 \cos 2\epsilon - 2pq \cos \epsilon + p^2)\},$$

$$N_1 = \frac{(c^2 - \mathfrak{g}^2)}{k^4} \sin \epsilon\, \{q\,(q \cos \epsilon - p)\,(p^2 \cos 2\epsilon - 2pq \cos \epsilon + q^2) + 2p\,(p \cos \epsilon - q)\,(q^2 \cos 2\epsilon - 2pq \cos \epsilon + p^2)\},$$

$$P_1 = \frac{\sqrt{c^2 - \mathfrak{g}^2}}{k} \sin^2 \epsilon \,.\, p\,(3pq^2 - p^3 - 2q^3 \cos \epsilon),$$

$$Q_1 = \sin^3 \epsilon \,.\, p^2 q\,(q \cos \epsilon - p).$$

The coefficient of sin 2 ψ

$$= \tfrac{15}{4} ma^2b^2 \left\{ -\left(\frac{\sqrt{c^2-\mathfrak{g}^2}}{k} (q \sin 2\epsilon - p \sin \epsilon) \right.\right.$$

$$\left. + (q \cos 2\epsilon - p \cos \epsilon)\, t \right) \frac{1}{(c^2+k^2t^2)^{\frac{7}{2}}} + 7 \left(\frac{(c^2-\mathfrak{g}^2)}{k^2} (p \cos \epsilon - q)(p - q \cos \epsilon) \right.$$

$$\left.\left. + \frac{\sqrt{c^2-\mathfrak{g}^2}}{k} \sin \epsilon\, (p^2 - q^2)\, t + pq \sin^2 \epsilon\, t^2 \right) \frac{1}{(c^2+k^2t^2)^{\frac{9}{2}}} \right\} \ldots\ldots (66).$$

§ 23. To find the effect of the vortex AB on CD we require the expressions for the velocity perpendicular to the plane of the vortex ring CD and along its radius vector.

The velocity perpendicular to the plane of CD

$$= w \cos \epsilon + u \sin \epsilon.$$

Now in this expression, the most important terms are the coefficients of cos ψ and sin ψ, because these terms, as we shall see, determine the deflection of the vortex. We shall therefore proceed to find these terms first.

The coefficient of cos ψ in the expression for the velocity perpendicular to the plane of CD may be written as

$$\tfrac{3}{2} \frac{ma^2b}{(c^2+k^2t^2)^{\frac{7}{2}}} (A + Bt + Ct^2 + Dt^3),$$

where

$$A = \frac{\sqrt{(c^2-\mathfrak{g}^2)}}{k} \left\{ c^2 (p \cos 2\epsilon - q \cos \epsilon) - 5 \sin^2 \epsilon \frac{(c^2-\mathfrak{g}^2)}{k^2} pq (p \cos \epsilon - q) \right\},$$

$$B = c^2 (p \sin 2\epsilon - q \sin \epsilon)$$

$$+ \frac{5(c^2-\mathfrak{g}^2)}{k^2} \{ q^3 - q^2 p \cos \epsilon - qp^2 (1 + \sin^2 \epsilon) + p^3 \cos \epsilon \},$$

$$C = \frac{\sqrt{c^2-\mathfrak{g}^2}}{k} \{ 4k^2 (q \cos \epsilon - p) + \sin^2 \epsilon \,.\, p\, (8p^2 - 7q^2 - pq \cos \epsilon) \},$$

$$D = \sin \epsilon \{ 5p^2 q \sin^2 \epsilon - k^2 (q + 3p \cos \epsilon) \}.$$

The coefficient of sin ψ

$$= \tfrac{3}{2} \frac{ma^2b\mathfrak{g}}{(c^2+k^2t^2)^{\frac{7}{2}}} (A' + B't + C't^2) \ldots\ldots\ldots\ldots (67),$$

where

$$A' = c^2 \cos \epsilon - \frac{5(c^2-\mathfrak{g}^2)}{k^2} pq \sin^2 \epsilon,$$

$$B' = \frac{5\sqrt{(c^2-\mathfrak{g}^2)}}{k} \{ 2pq \sin \epsilon \cos \epsilon - 2q^2 \sin \epsilon + k^2 \},$$

$$C' = k^2 \cos \epsilon - 5 (q \cos \epsilon - p)(q - p \cos \epsilon).$$

Now, since the equation to the vortex CD is

$$z' = \mathfrak{z}' + \Sigma\,(\gamma'_n \cos n\psi + \delta'_n \sin n\psi).$$

The velocity perpendicular to the plane of the vortex

$$= \frac{d\mathfrak{z}'}{dt} + \Sigma\left(\frac{d\gamma'_n}{dt}\cos n\psi + \frac{d\delta'_n}{dt}\sin n\psi\right),$$

since as δ'_n, γ'_n and Ψ are all small quantities we may neglect

$$n\,(\delta'_n \cos n\psi - \gamma'_n \sin n\psi)\,\Psi.$$

Thus $\dfrac{d\gamma'_1}{dt}$ = coefficient of $\cos\psi$ in the expression for the velocity perpendicular to the plane of the vortex CD.

A reference to equation (43) will shew that the vortex CD contributes nothing to this term, so that

$$\frac{d\gamma'_1}{dt} = \tfrac{3}{2}\,\frac{ma^2b}{(c^2+k^2t^2)^{\frac{7}{2}}}\,(A + Bt + Ct^2 + Dt^3).$$

Integrating we find

$$\gamma'_1 = \tfrac{3}{2}\,ma^2b\left\{\tfrac{1}{5}\,\frac{Dc^2/k^4 - B/k^2}{(c^2+k^2t^2)^{\frac{5}{2}}} - \tfrac{1}{3}\,\frac{D/k^4}{(c^2+k^2t^2)^{\frac{3}{2}}} + \tfrac{1}{5}\,\frac{A - Cc^2/k^2}{c^2}\,\frac{t}{(c^2+k^2t^2)^{\frac{5}{2}}}\right.$$
$$\left. + \tfrac{1}{15}\,\frac{\{4A/c^4 + C/c^2k^2\}\,t}{(c^2+k^2t^2)^{\frac{3}{2}}} + \tfrac{1}{15}\left(\frac{8A}{c^6} + \frac{2C}{c^4k^2}\right)\left(\frac{t}{(c^2+k^2t^2)^{\frac{1}{2}}} + \frac{1}{k}\right)\right\} \quad\text{.......(68)},$$

where the arbitrary constant arising from the integration has been determined so as to make $\gamma'_1 = 0$ when $t = -\infty$.

If we substitute for A, B, C, D the values given above, we shall get the value for γ'_1 at each instant of the collision; but at present we shall only consider the change in γ'_1 when it has got so far away from the vortex AB that its motion is again undisturbed. We can find this change in γ'_1 by putting $t = \infty$ in the above formula, on doing this we find

$$\gamma'_1 = \tfrac{1}{5}\,ma^2b\left(\frac{4A}{c^6} + \frac{C}{c^4k^2}\right)\frac{2}{k};$$

or substituting for A and C their values,

$$\gamma'_1 = \frac{2ma^2bpq}{c^4k^4}\,(q - p\cos\epsilon)\left(1 - \frac{4\mathfrak{g}^2}{c^2}\right)\sqrt{(c^2 - \mathfrak{g}^2)}\,\sin^2\epsilon \quad\text{....(69)}.$$

§ 24. We have similarly

$\dfrac{d\delta'_1}{dt}$ = coefficient of $\sin\psi$ in the expression for the velocity perpendicular to the plane of the vortex

$$= \tfrac{3}{2}\,\frac{ma^2b\mathfrak{g}}{(c^2+k^2t^2)^{\frac{7}{2}}}\,(A' + B't + C't^2).$$

Integrating we find

$$\delta'_1 = \tfrac{8}{3}\, ma^2 b\mathfrak{g} \left\{ -\tfrac{1}{3}\frac{B'/k^2}{(c^2+k^2t^2)^{\frac{3}{2}}} + \tfrac{1}{5}\frac{(A'-C'c^2/k^2)}{c^2}\frac{t}{(c^2+k^2t^2)^{\frac{5}{2}}} \right.$$

$$\left. + \tfrac{1}{15}\frac{(4A'/c^4+C'/c^2k^2)t}{(c^2+k^2t^2)^{\frac{3}{2}}} + \tfrac{1}{15}\left(\frac{8A'}{c^6}+\frac{2C'}{c^4k^2}\right)\left(\frac{t}{(c^2+k^2t^2)^{\frac{1}{2}}}+\frac{1}{k}\right)\right\} \text{......(70)},$$

where the arbitrary constant arising from the integration has been determined so as to make $\delta'_1 = 0$ when $t = -\infty$. The change in δ'_1 when the vortex CD is so far away from AB that its motion is undisturbed is given by

$$\delta'_1 = \tfrac{8}{3}\, ma^2 b\mathfrak{g}\, \tfrac{1}{15}\left(\frac{8A'}{c^6}+\frac{2C'}{c^4k^2}\right)\frac{2}{k}.$$

Substituting we find

$$\delta'_1 = -\frac{2ma^2b\mathfrak{g}}{c^4k^3}\sin^2\epsilon\,.\,pq\left(1-\frac{4\mathfrak{g}^2}{3c^2}\right) \text{...............(71).}$$

§ 25. We have in paragraph (6) investigated the changes in the direction cosines of the direction of motion of the vortex ring due to changes in the coefficients γ'_1 and δ'_1. From that investigation we find that the direction cosines of the direction of motion of the vortex CD after the impact are

$$\sin\epsilon - \frac{\gamma'_1}{b}\cos\epsilon,$$

$$-\frac{\delta'_1}{b},$$

$$\cos\epsilon + \frac{\gamma'_1}{b}\sin\epsilon,$$

or substituting for γ'_1 and δ'_1 the values just found, the direction cosines become

$$\sin\epsilon - \frac{2ma^2}{c^4k^4}\sqrt{(c^2-\mathfrak{g}^2)}\left(1-\frac{4\mathfrak{g}^2}{c^2}\right)\sin^2\epsilon\cos\epsilon.\,pq\,(q-p\cos\epsilon),$$

$$\frac{2ma^2\mathfrak{g}}{c^4k^3}\sin^2\epsilon\,.\,pq\left(1-\frac{4\mathfrak{g}^2}{3c^2}\right),$$

$$\cos\epsilon + \frac{2ma^2}{c^4k^4}\sqrt{(c^2-\mathfrak{g}^2)}\left(1-\frac{4\mathfrak{g}^2}{c^2}\right)\sin^3\epsilon.\,pq\,(q-p\cos\epsilon).$$

Thus if A, B, C (fig. 5) be the points where the axes of x, y, z cut a sphere with the origin for centre and P the point where a parallel through this centre to the direction of motion of the vortex CD before the collision cuts the sphere.

Then if the vortex CD be the first to intersect the shortest distance between the directions of motion of the vortices, P' will be the point where a parallel to the direction of motion after impact

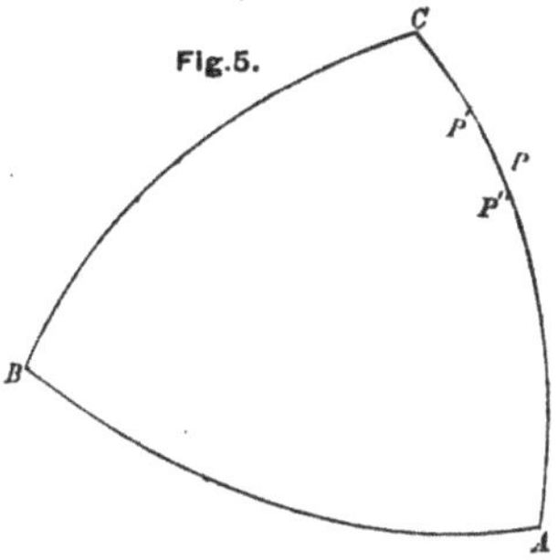

Fig. 5.

cuts the sphere, supposing $\mathfrak{g}$ to be positive and $< \frac{1}{2}c$ and the velocity of CD greater than the velocity of AB resolved along the direction of motion of CD, *i.e.* if $q - p\cos\epsilon$ be positive. We may describe this by saying that the direction of motion of the vortex ring is altered in the same way as it would be if the vortex ring received an impulse parallel to the shortest distance between the directions of motion of the vortices and another impulse perpendicular both to its own direction of motion and the shortest distance; the first impulse being from and the second towards the vortex AB. In this case the angle between the direction of motion of CD and the original direction of motion of AB is diminished by the impact.

If the vortex AB be the first to intersect the shortest distance then we must change the sign of $\sqrt{(c^2 - \mathfrak{g}^2)}$ in the expressions for $\mathfrak{f}$ and $\mathfrak{h}$; this will change the sign of γ'_1 but will leave δ'_1 unaltered, and consequently P'' the point where the direction of motion of CD after the impact intersects the sphere of reference will be situated as in the figure; in this case the angle between the direction of motion of CD and the original direction of motion of AB is increased by the impact. The angle through which the direction of motion of CD is deflected

$$= \sqrt{\left(\frac{\gamma_1^2}{b^2} + \frac{\delta_1^2}{b^2}\right)}$$

$$= \frac{2ma^2pq}{c^4k^3}\sin^2\epsilon\left\{\frac{(q - p\cos\epsilon)^2}{k^2}\left(1 - \frac{4\mathfrak{g}^2}{c^2}\right)^2(c^2 - \mathfrak{g}^2) + \mathfrak{g}^2\left(1 - \frac{4\mathfrak{g}^2}{3c^2}\right)^2\right\}^{\frac{1}{2}} \quad \text{.....................(72).}$$

If the paths of the vortices intersect so that $\mathfrak{g} = 0$, this becomes

$$\frac{2ma^2\sin^2\epsilon}{c^3k^4}pq(q - p\cos\epsilon) \quad \text{.....................(73),}$$

or the deflection is *cœteris paribus* inversely proportional to the cube of the shortest distance between the vortices.

If the paths of the vortices do not intersect, but the vortices move so as to come as close together as possible, then $c = \mathfrak{g}$, and the deflection

$$= \frac{2ma^2 \sin^2\epsilon \,.\, pq}{3c^3k^3} \quad \ldots\ldots\ldots\ldots\ldots\ldots\ldots(74).$$

This is again inversely proportional to the cube of the distance.

If in the two cases above, c be the same, then the deflection when the paths of the vortices intersect will be greater, equal to or less than when they do not, according as $8\,(q - p\cos\epsilon)^2$ is greater, equal to, or less than $p^2 \sin^2 \epsilon$; thus, unless the relative velocity of the vortices perpendicular to the direction of motion of *CD* is great compared with that along *CD*, the deflection will be greater when the directions of motion of the vortices intersect than when they do not.

The expression for the deflection simplifies when the line joining the vortices at the instant when they are nearest together is inclined at an angle of 30^0 to the shortest distance between their directions of motion, in this case $\mathfrak{g} = c \cos 30^0 = c\,\tfrac{1}{2}\sqrt{3}$, thus $\delta'_1 = 0$ as $1 - \dfrac{4\mathfrak{g}^2}{3c^2}$ vanishes, and the deflection

$$= \frac{2ma^2 \sin^2\epsilon \,.\, pq\,(q - p\cos\epsilon)}{c^3k^4},$$

which, if c be the same, is the same as when the vortices intersect.

§ 26. We have next to consider how the vortex *CD* is altered in size by the collision.

We know that if α'_0 be the alteration in the radius of the vortex *CD* that

$\dfrac{d\alpha'_0}{dt}$ = coefficient of the term independent of ψ in the expression for the velocity along the radius vector of *CD*.

Now a reference to equation (38) will shew that the vortex *CD* contributes nothing to this term itself, so that

$\dfrac{d\alpha'_0}{dt}$ = coefficient of the term independent of ψ in the expression for the velocity along the radius vector of *CD* due to the vortex *AB*.

Since λ, μ, ν, the direction-cosines of a radius vector, are by § 6 given by the equations

$$\lambda = \cos\epsilon\cos\psi,$$
$$\mu = \sin\psi,$$
$$\nu = -\sin\epsilon\cos\psi,$$

$\frac{d\alpha'_0}{dt}$ = coefficient of the term independent of ψ in

$$u\cos\epsilon\cos\psi + v\sin\psi - w\sin\epsilon\cos\psi.$$

Hence by equations (53), (59), (63),

$$\frac{d\alpha'_0}{dt} = \tfrac{3}{4}\frac{ma^2b}{(c^2+k^2t^2)^{\frac{7}{2}}}(F + Gt + Ht^2 + Kt^3),$$

where

$$F = \frac{\sqrt{(c^2-\mathfrak{g}^2)}}{k^3}\sin\epsilon\,[c^2\{p^2q(4-\cos^2\epsilon) - 2p^3\cos\epsilon - q^3\} - 5\mathfrak{g}^2\sin^2\epsilon\,.\,p^2q],$$

$$G = c^2\left\{(q\cos\epsilon - p)\left(2 - \frac{5(q-p\cos\epsilon)^2}{k^2}\right) + \sin^2\epsilon\,.\,p\left(3 - \frac{5(p^2-q^2)}{k^2}\right)\right\},$$

$$H = \frac{\sqrt{(c^2-\mathfrak{g}^2)}}{k}\sin\epsilon\,(8p^3\cos\epsilon - p^2q\cos^2\epsilon - 11p^2q + 4q^3),$$

$$K = k^2\{2(q\cos\epsilon - p) + 3p\sin^2\epsilon\} - 5pq\sin^2\epsilon\,(q - p\cos\epsilon).$$

Integrating, we find

$$\alpha'_0 = \tfrac{3}{4}ma^2b\left\{\tfrac{1}{5}\frac{Kc^2/k^4 - G/k^2}{(c^2+k^2t^2)^{\frac{5}{2}}} - \tfrac{1}{3}\frac{K/k^4}{(c^2+k^2t^2)^{\frac{3}{2}}} + \tfrac{1}{5}\frac{F - Hc^2/k^2}{c^2}\frac{t}{(c^2+k^2t^2)^{\frac{5}{2}}}\right.$$
$$\left. + \tfrac{1}{15}\frac{(4F/c^4 + H/c^2k^2)\,t}{(c^2+k^2t^2)^{\frac{3}{2}}} + \tfrac{1}{15}\left(\frac{8F}{c^6} + \frac{2H}{c^4k^2}\right)\left(\frac{t}{(c^2+k^2t^2)^{\frac{1}{2}}} + \frac{1}{k}\right)\right\},$$

where the arbitrary constant arising from the integration has been determined so as to make $\alpha'_0 = 0$ when $t = -\infty$. If we substitute for F, G, H, K the values just written we shall get the change in the radius at any instant, but at present we shall only consider the change in the radius of CD when it has got so far away from the vortex AB that its motion is again undisturbed. We can find this change in the radius by putting $t = \infty$ in the above formula; doing this we find

$$\alpha'_0 = \frac{ma^2b}{5k}\left(\frac{4F}{c^6} + \frac{H}{c^4k^2}\right).$$

Substituting for F and H their values, we find

$$\alpha'_0 = \frac{ma^2b\sin^3\epsilon\,.\,p^2q}{k^3c^4}\left(1 - \frac{4\mathfrak{g}^2}{c^2}\right)\sqrt{(c^2-\mathfrak{g}^2)} \quad\text{........(74*)}.$$

Thus we see that the radius of the vortex which first passes through the shortest distance between their directions of motion is increased, provided $c > 2\mathfrak{g}$. If AB had first intersected the shortest distance we should have had to change the sign of $\sqrt{(c^2 - \mathfrak{g}^2)}$, then α'_0 would be negative, and the radius of CD would be diminished.

If the directions of motion of the vortices intersect, so that $\mathfrak{g} = 0$, then

$$\alpha'_0 = \frac{ma^2 b \sin^2 \epsilon \, . \, p^2 q}{k^3 c^3},$$

or the increase in radius is *cæteris paribus* inversely proportional to the cube of the shortest distance between the vortices.

If the directions of motion of the vortices do not intersect, but the vortices move so as to come as close together as possible, then $c = \mathfrak{g}$, and $\alpha'_0 = 0$, and the radius of the vortex in this case is not altered by the collision.

If $c = 2\mathfrak{g}$, or if the line joining the vortices when they are nearest together be inclined at an angle of 60° to the shortest distance between the directions of motion of the vortices, then $\alpha'_0 = 0$, or in this case again the radius of the vortex is not altered by the collision. Thus we see for our present purpose we may divide collisions into two classes. In the first class the line joining the centres of the vortices when they are nearest together is inclined at an angle greater than 60° to the shortest distance between the directions of motion of the vortices. In this case the vortex which first passes through the shortest distance increases in radius, and consequently decreases in velocity and increases in energy, while the other vortex decreases in radius and energy and increases in velocity.

In the second class of collisions the line joining the centres of the vortices when they are nearest together is inclined at an angle less than 60° to the shortest distance between the directions of motion of the vortices. In this case the vortex which first passes through the shortest distance decreases in radius, and consequently increases in velocity and decreases in energy, while the other vortex increases in radius and energy and decreases in velocity.

§ 27. Having found the change in the radius and the change in the direction of motion of the vortex, we can find the changes in the components of the momentum of the vortex referred to any axes.

Let $\mathfrak{I}'$ be the momentum of the vortex CD; $\mathfrak{P}'$, $\mathfrak{Q}'$, $\mathfrak{R}'$ its components parallel to the axes of x, y, z respectively, l', m', n' the direction-cosines of the normal to the plane of the vortex.

Thus $$\mathfrak{I}' = m'\pi b^2, \quad \mathfrak{P}' = m'\pi b^2 l',$$

so $$\delta\mathfrak{P}' = 2\pi m' b\delta b l' + m'\pi b^2 \delta l'$$
$$= 2\frac{\alpha'_0}{b}\mathfrak{P}' + \mathfrak{I}'\delta l',$$

similarly,
$$\delta\mathfrak{Q}' = 2\frac{\alpha'_0}{b}\mathfrak{Q}' + \mathfrak{L}'\delta m',$$
$$\delta\mathfrak{R}' = 2\frac{\alpha'_0}{b}\mathfrak{R}' + \mathfrak{L}'\delta n'.$$

It remains to find $\delta l'$, $\delta m'$, $\delta n'$ in terms of γ'_0 and δ'_0. Now if $\mathfrak{L}$, $\mathfrak{P}$, $\mathfrak{Q}$, $\mathfrak{R}$ denote the same quantities for the vortex AB as the same letters accented do for the vortex CD, then it is easy to prove that the direction-cosines of the old axes referred to the new are as follows.

The direction-cosines of the old axis of x are
$$\frac{\mathfrak{P}'\mathfrak{L} - \mathfrak{P}\mathfrak{L}'\cos\epsilon}{\mathfrak{L}.\mathfrak{L}'\sin\epsilon},\ \frac{\mathfrak{Q}'\mathfrak{L} - \mathfrak{Q}\mathfrak{L}'\cos\epsilon}{\mathfrak{L}.\mathfrak{L}'\sin\epsilon},\ \frac{\mathfrak{R}'\mathfrak{L} - \mathfrak{R}\mathfrak{L}'\cos\epsilon}{\mathfrak{L}.\mathfrak{L}'\sin\epsilon}.$$
The direction-cosines of the old axis of y are
$$\frac{\mathfrak{Q}\mathfrak{R}' - \mathfrak{R}\mathfrak{Q}'}{\mathfrak{L}.\mathfrak{L}'\sin\epsilon},\ \frac{\mathfrak{R}\mathfrak{P}' - \mathfrak{P}\mathfrak{R}'}{\mathfrak{L}.\mathfrak{L}'\sin\epsilon},\ \frac{\mathfrak{P}\mathfrak{Q}' - \mathfrak{Q}\mathfrak{P}'}{\mathfrak{L}.\mathfrak{L}'\sin\epsilon}.$$
The direction-cosines of the old axis of z are
$$\frac{\mathfrak{P}}{\mathfrak{L}},\ \frac{\mathfrak{Q}}{\mathfrak{L}},\ \frac{\mathfrak{R}}{\mathfrak{L}}.$$

Thus if λ, μ, ν be the direction-cosines of the normal to the plane of the vortex CD referred to the old axes, then
$$\delta l' = \frac{\delta\lambda\,(\mathfrak{P}'\mathfrak{L} - \mathfrak{P}\mathfrak{L}'\cos\epsilon)}{\mathfrak{L}.\mathfrak{L}'\sin\epsilon} + \frac{\delta\mu\,(\mathfrak{Q}\mathfrak{R}' - \mathfrak{R}\mathfrak{Q}')}{\mathfrak{L}.\mathfrak{L}'\sin\epsilon} + \frac{\delta\nu.\mathfrak{P}}{\mathfrak{L}},$$
with symmetrical expressions for $\delta m'$ and $\delta n'$.

Now by § 6
$$\delta\lambda = -\frac{\gamma'_1}{b}\cos\epsilon,$$
$$\delta\mu = -\frac{\delta'_1}{b},$$
$$\delta\nu = \frac{\gamma'_1}{b}\sin\epsilon.$$

Substituting for γ'_1 and δ'_1 their values, we find
$$\delta l' = \frac{2ma^2pq\sin\epsilon}{c^4k^2\mathfrak{L}.\mathfrak{L}'}\left\{\frac{q - p\cos\epsilon}{k}\sqrt{(c^2 - \mathfrak{g}^2)}\left(1 - \frac{4\mathfrak{g}^2}{c^2}\right)(\mathfrak{P}\mathfrak{L}' - \mathfrak{L}\mathfrak{P}'\cos\epsilon)\right.$$
$$\left. + \mathfrak{g}\left(1 - \frac{4\mathfrak{g}^2}{3c^2}\right)(\mathfrak{Q}\mathfrak{R}' - \mathfrak{R}\mathfrak{Q}')\right\},$$
with symmetrical expressions for $\delta m'$ and $\delta n'$.

Thus

$$\delta\mathfrak{P}' = \frac{2ma^2pq\sin\epsilon}{c^4k^3\mathfrak{I}\,.\,\mathfrak{I}'}\left\{\frac{\sqrt{(c^2-\mathfrak{g}^2)}}{k}\left(1-\frac{4\mathfrak{g}^2}{c^2}\right)\right.$$

$$\left.\times\,\mathfrak{I}'\{\mathfrak{P}\mathfrak{I}'(q-p\cos\epsilon)-\mathfrak{P}'\mathfrak{I}(q\cos\epsilon-p)\}+\mathfrak{g}\left(1-\frac{4\mathfrak{g}^2}{3c^2}\right)\mathfrak{I}'(\mathfrak{Q}\mathfrak{R}'-\mathfrak{R}\mathfrak{Q}')\right\}$$

$$=\frac{pq\sin\epsilon}{\pi c^4k^3}\left\{\frac{\sqrt{(c^2-\mathfrak{g}^2)}}{k}\left(1-\frac{4\mathfrak{g}^2}{c^2}\right)\right.$$

$$\left.\times\{\mathfrak{P}\mathfrak{I}'(q-p\cos\epsilon)-\mathfrak{P}'\mathfrak{I}(q\cos\epsilon-p)\}+\mathfrak{g}\left(1-\frac{4\mathfrak{g}^2}{3c^2}\right)(\mathfrak{Q}\mathfrak{R}'-\mathfrak{R}\mathfrak{Q}')\right\}$$

.........(75),

with symmetrical expressions for $\delta\mathfrak{Q}'$ and $\delta\mathfrak{R}'$.

If ϕ be the angle which the line joining the centres of the vortices when they are nearest together makes with the shortest distance between the paths of the centres of the vortex rings, then

$$\mathfrak{g} = c\cos\phi,$$

so $$\left(1-4\frac{\mathfrak{g}^2}{c^2}\right)\sqrt{c^2-\mathfrak{g}^2} = c\sin\phi\,(4\sin^2\phi-3) = -c\sin 3\phi,$$

and $$\mathfrak{g}\left(1-\frac{4\mathfrak{g}^2}{3c^2}\right) = c\cos\phi\left(1-\frac{4\cos^2\phi}{3}\right) = -\frac{c}{3}\cos 3\phi.$$

Thus

$$\delta\mathfrak{P}' = -\frac{4pq\sin\epsilon}{\pi c^3k^4}\Big[\sin 3\phi\,\{\mathfrak{P}\mathfrak{I}'(q-p\cos\epsilon)-\mathfrak{P}\mathfrak{I}(q\cos\epsilon-p)\}$$

$$-\tfrac{1}{3}\cos 3\phi\,\mathfrak{R}\,(\mathfrak{Q}\mathfrak{R}'-\mathfrak{R}\mathfrak{Q}')\Big],$$

with symmetrical expressions for $\delta\mathfrak{Q}'$ and $\delta\mathfrak{R}'$.

Since $\mathfrak{P}+\mathfrak{P}'$ is constant throughout the motion

$$\delta\mathfrak{P} = -\delta\mathfrak{P}',$$

similarly $$\delta\mathfrak{Q} = -\delta\mathfrak{Q}',$$

$$\delta\mathfrak{R} = -\delta\mathfrak{R}'.$$

§ 28. We can now sum up the effects of the collision upon the vortex rings AB and CD. We shall find it convenient to express them in terms of the angle ϕ used in the last paragraph: ϕ is the angle which the line joining the centres of the vortex rings when they are nearest together makes with the shortest distance between the paths of the centres of the vortex rings, ϕ is positive for the vortex ring which first intersects the shortest distance between the paths, negative for the other ring, so that with a given $\mathfrak{g}$, ϕ may be regarded as giving the delay of one vortex behind the other.

§ 29. Let us first consider the effect of the collision on the radii of the vortex rings.

The radius of the vortex ring CD is diminished by

$$\frac{ma^2b}{c^3k^4}p^2q\sin^3\epsilon\sin 3\phi.$$

Thus the radius of the ring is diminished or increased according as $\sin 3\phi$ is positive or negative. Now ϕ is positive for one vortex ring negative for the other, thus $\sin 3\phi$ is positive for one vortex ring negative for the other, so that if the radius of one vortex ring is increased by the collision the radius of the other will be diminished. When ϕ is less than 60° the vortex ring which first passes through the shortest distance between the paths of the centres of the rings diminishes in radius and the other one increases. When ϕ is greater than 60° the vortex ring which first passes through the shortest distance between the paths increases in radius and the other one diminishes. When the paths of the centres of the vortex rings intersect ϕ is 90°, so that the vortex ring which first passes through the shortest distance, which in this case is the point of intersection of the paths, is the one which increases in radius. When ϕ is zero or the vortex rings intersect the shortest distance simultaneously there is no change in the radius of either vortex ring, and this is also the case when ϕ is 60°.

§ 30. Let us now consider the bending of the path of the centre of one of the vortex rings perpendicular to the plane through the centre of the other ring and parallel to the original paths of both the vortex rings.

We see by equation (71) that the path of the centre of the vortex ring CD is bent towards this plane through an angle

$$\tfrac{2}{3}\frac{ma^2}{c^3k^3}pq\sin^2\epsilon\cos 3\phi\,;$$

this does not change sign with ϕ, and whichever vortex first passes through the shortest distance the deflection is given by the rule that the path of a vortex ring is bent towards or from the plane through the centre of the other vortex and parallel to the original directions of both vortices according as $\cos 3\phi$ is positive or negative, so that if ϕ is less than 30° the path of the vortex is bent towards, and if ϕ be greater than 30° from this plane. It follows from this expression for the deflection that if we have a large quantity of vortex rings uniformly distributed they will on the whole repel a vortex ring passing by them.

§ 31. Let us now consider the bending of the paths of the vortices in the plane parallel to the original paths of both vortex rings. Equation (69) shews that the path of the vortex ring CD is bent in this plane through an angle

$$-\frac{2ma^2}{c^3k^4}\sin^2\epsilon\sin 3\phi\, pq\,(q-p\cos\epsilon)$$

towards the direction of motion of the other vortex. Thus the direction of motion of one vortex is bent from or towards the direction of motion of the other according as $\sin 3\phi\,(q - p\cos\epsilon)$ is positive or negative. Comparing this result with the result for the change in the radius, we see that if the velocity of a vortex ring CD be greater than the velocity of the other vortex AB resolved along the direction of motion of CD, then the path of each vortex will be bent towards the direction of motion of the other when its radius is increased and away from the direction of motion of the other when its radius is diminished, while if the velocity of the vortex be less than the velocity of the other resolved along its direction of motion, the direction of motion will be bent from the direction of the other when its radius is increased and *vice versâ*. The rules for finding the alteration in the radius were given before.

§ 32. Equation (75) shews that the effect of the collision is the same as if an impulse

$$-\frac{pq\mathfrak{I}\,.\,\mathfrak{I}'}{\pi\rho c^3k^3}\sin^2\epsilon\sin 3\phi,$$

parallel to the resultant of velocities $p - q\cos\epsilon$ and $q - p\cos\epsilon$ along the paths of vortices (CD) and (AB) respectively, and an impulse

$$-\frac{pq\mathfrak{I}\,.\,\mathfrak{I}'}{3\pi\rho c^2k^3}\sin^2\epsilon\cos 3\phi,$$

parallel to the shortest distance between the original paths of the vortex rings, were given to one of the vortices and equal and opposite impulses to the other; here $\mathfrak{I}$ and $\mathfrak{I}'$ are the momenta of the vortices.

§ 33. We have so far been engaged with the changes in the magnitude and position of the vortex ring CD, and have not considered the changes in shape which the vortex ring suffers from the collision. These changes will be expressed by the quantities α_2, β_2, α_3, β_3, &c. We must now investigate the values of these quantities.

Now we know

$\frac{d\alpha'_2}{dt}$ = coefficient of $\cos 2\psi$ in the expression for the velocity along the radius vector.

A reference to equation (38) will shew that the vortex ring CD itself contributes to this coefficient the term

$$-\frac{2m'}{\pi b^2}\log\frac{8b}{e'}\,.\,\gamma'_2.$$

The vortex ring AB contributes, as we see from equations (53), (59), and (63), a term

$$\tfrac{3}{4}\frac{ma^2b}{(c^2+k^2t^2)^{\frac{7}{2}}}\{F'+G't+H't^2+K't^3),$$

where

$$F' = \frac{c^3}{k^3}\sin\epsilon\,\{p^2q\,(2-\cos^2\epsilon)+4pq^2\cos\epsilon-3q^3-2p^3\cos\epsilon\},$$

$$G' = c^2\left\{p\sin^2\epsilon\left(3-5\,\frac{(p^2-q^2)}{k^2}\right)-5\,(q\cos\epsilon-p)\,\frac{(q-p\cos\epsilon)^2}{k^2}\right\},$$

$$H' = \frac{c\sin\epsilon}{k}\,\{8p^3\cos\epsilon-p^2q\,(2+\cos^2\epsilon)+pq^2\,(4\cos\epsilon-11)+2q^3\},$$

$$K' = 3k^2p\sin^2\epsilon-5pq\sin^2\epsilon\,(q-p\cos\epsilon),$$

where, in order to make the work as simple as possible, we have put $\mathfrak{g}=0$; so that the undisturbed paths of the vortices intersect.

Thus

$$\frac{d\alpha'_2}{dt} = -\frac{2m'}{\pi b^2}\log\frac{8b}{e'}\,.\,\gamma'_2+\tfrac{3}{4}\,\frac{ma^2b}{(c^2+k^2t^2)^{\frac{7}{2}}}(F'+G't+H't^2+K't^3),$$

say

$$\frac{d\alpha'_2}{dt} = -\frac{2m'}{\pi b^2}\log\frac{8b}{e'}\,.\,\gamma'_2+f(t).$$

Now $\frac{d\gamma'_2}{dt}$ = the coefficient of $\cos 2\psi$ in the expression for the velocity perpendicular to the plane of the vortex CD.

The vortex CD itself contributes to this coefficient the term

$$\tfrac{3}{2}\frac{m'}{\pi b^2}\log\frac{8b}{e'}\,.\,\alpha'_2.$$

The vortex AB contributes, as we see from equations (55) and (65), the term

$$\tfrac{3}{2}\,\frac{ma^2b}{(c^2+k^2t^2)^{\frac{9}{2}}}(F''+G''t+H''t^2+K''t^3+L''t^4).$$

Say for brevity $F(t)$, where if, as before, we put $\mathfrak{g}=0$,

$$F'' = \frac{c^4}{k^4}(p\cos\epsilon-q)\,\{\tfrac{35}{4}(p\cos\epsilon-q)\,(3pq\sin^2\epsilon-k^2\cos\epsilon)$$
$$-5k^2(\tfrac{1}{2}p\sin^2\epsilon+p-q\cos\epsilon)\},$$

$$G'' = \frac{5c^3\sin\epsilon}{k^3}\,[(q^2-p^2)\,\{\tfrac{21}{4}(p\cos\epsilon-q)^2+\tfrac{3}{2}k^2\}$$
$$+(p\cos\epsilon-q)\,p\,(\tfrac{21}{4}pq\sin^2\epsilon-\tfrac{5}{2}k^2)],$$

$$H'' = \frac{pc^2 \sin^2 \epsilon}{k^2} [\{p^2 q \sin^2 \epsilon + 2 (p \cos \epsilon - q)(p^2 - q^2) - (p \cos \epsilon - q)^3\} (q \cos \epsilon - p) - \tfrac{1}{2} 5k^2 \{\tfrac{7}{2} \cos \epsilon . k^2 + (p^2 + q^2) \cos \epsilon - 2pq\}],$$

$$K'' = \tfrac{5}{4} \frac{pc \sin \epsilon}{k} \{21 p \sin^2 \epsilon (q^2 - p^2) - k^2 (5p \cos \epsilon + 6q)\},$$

$$L'' = \tfrac{5}{4} p \sin^2 \epsilon \{21 p (q - p \cos \epsilon)(q \cos \epsilon - p) - k^2 (5p \cos \epsilon + 6q)\}.$$

Thus

$$\frac{d\gamma'_2}{dt} = \tfrac{3}{2} \frac{m'}{\pi b^2} \log \frac{8b}{e'} . \alpha'_2 + F(t);$$

differentiating this equation, and substituting for $\frac{d\alpha'_2}{dt}$ from the other equation, we find

$$\frac{d^2\gamma'_2}{dt^2} + 3 \left(\frac{m'}{\pi b^2} \log \frac{2b}{e'}\right)^2 \gamma'_2 = F'(t) + \tfrac{3}{2} \frac{m'}{b^2} \log \frac{2b}{e'} f(t)$$

$$= \chi(t) \text{ say};$$

or writing n^2 for $3 \left(\frac{m'}{\pi b^2} \log \frac{2b}{e'}\right)^2$,

$$\frac{d^2\gamma'_2}{dt^2} + n^2 \gamma'_2 = \chi(t).$$

The solution of this differential equation is

$$\gamma'_2 = A \cos nt + B \sin nt + \frac{\cos nt}{n} \int^t \chi(t') \sin nt' \, dt' - \frac{\sin nt}{n} \int^t \chi(t') \cos nt' \, dt',$$

or choosing the arbitary constants so that γ'_2 and $\frac{d\gamma'_2}{dt}$ both vanish when $t = -\infty$, we find

$$\gamma'_2 = \frac{\cos nt}{n} \int_{-\infty}^t \chi(t') \sin nt' \, dt' - \frac{\sin nt}{n} \int_{-\infty}^t \chi(t') \cos nt' \, dt'.$$

The integral

$$\int_{-\infty}^t \chi(t') \cos nt' \, dt'$$

involves integrals of the form

$$\int_{-\infty}^t \frac{\cos nt' \, dt'}{(c^2 + k^2 t'^2)^{\frac{1}{2}(2p+1)}} \text{ and } \int_{-\infty}^t \frac{t' \cos nt' \, dt'}{(c^2 + k^2 t'^2)^{\frac{1}{2}(2p+1)}}.$$

I have not been able to evaluate these integrals except when $t = \infty$.

In the expression for γ'_2 the terms under the integral express the effect of the vortex AB on CD. Now the vortex AB will only exert an appreciable effect on CD during the time the vortices are in the neighbourhood of the place where they are nearest together; and thus, after the collision, we may, without appreciable error, write the equation for γ'_2 as

$$\gamma'_2 = \frac{P\cos nt}{n} - \frac{Q\sin nt}{n},$$

where
$$P = \int_{-\infty}^{+\infty} \sin nt \,.\, \chi(t)\, dt,$$

$$Q = \int_{-\infty}^{+\infty} \cos nt \,.\, \chi(t)\, dt.$$

Thus the vortex rings are thrown by the collision into vibration, and after the collision is over the period of the vibration is $\frac{2\pi}{n}$, the same as the period of the corresponding free vibration of the vortex CD.

To find P and Q we have to find

$$\int_{-\infty}^{+\infty} \frac{\cos nt \,.\, dt}{(c^2 + k^2t^2)^{\frac{1}{2}(2p+1)}},$$

or if we write q for $\frac{c}{k}$,

$$\int_{-\infty}^{+\infty} \frac{\cos nt \,.\, dt}{(q^2 + t^2)^{\frac{1}{2}(2p+1)}}.$$

Now q is the time taken by the vortices to separate by a distance c, while $\frac{2\pi}{n}$ is (§ 13) of the same order as the time taken by the vortex CD to pass over a length equal to its diameter; but, since c is large compared with the diameter of the vortex, $\dfrac{q}{\frac{2\pi}{n}}$ or nq is large.

Let
$$\int_0^{\infty} \frac{\cos nt \,.\, dt}{(q^2 + t^2)^{\frac{1}{2}(2p+1)}} = n^p v_p.$$

By differentiation we find

$$v_{p+1} = -\frac{q^{-2}}{(2p+1)}\left(\frac{dv_p}{dn} - \frac{p}{n}v_p\right).$$

Hence we find

$$v_p = \frac{(-1)^p q^{-2p}}{1.3.5\ldots(2p-1)}\left(\frac{d}{dn} - \frac{p-1}{n}\right)\left(\frac{d}{dn} - \frac{p-2}{n}\right)\ldots\frac{d}{dn}v_0.$$

This may be written

$$v_p = \frac{(-2n)^p q^{-2p}}{1.3.5\ldots(2p-1)} \left(\frac{d}{d(n^2)}\right)^p v_0 \ldots\ldots\ldots\ldots\ldots\ldots(76).$$

We can easily verify that v_p satisfies the differential equation

$$\frac{d^2 v_p}{dn^2} + \frac{1}{n}\frac{dv_p}{dn} - \left(\frac{p^2}{n^2} + q^2\right) v_p = 0.$$

Let us assume

$$v_p = \frac{e^{-nq}}{n^x}\left(A_0 + \frac{A_1}{n} + \frac{A_2}{n^2} + \ldots\right).$$

If we substitute this expression for v_p in the differential equation, and equate to zero the various powers of n, we get the equations

$$A_0(2qx - q) = 0,$$

$$2q(x+1)A_1 + \quad x(x+1)A_0 \quad -qA_1 - \quad xA_0 \quad -p^2A_0 \quad = 0,$$

$$2q(x+2)A_2 + (x+1)(x+2)A_1 - qA_2 - (x+1)A_1 - p^2A_1 = 0,$$

....

$$2q(x+m)A_m + (x+m-1)(x+m)A_{m-1} - qA_m - (x+m-1)A_{m-1} - p^2A_{m-1} = 0;$$

these give

$$x = \tfrac{1}{2},$$

$$2qA_1 + \quad (\tfrac{1}{4} - p^2)A_0 = 0,$$

$$4qA_2 + \quad (\tfrac{9}{4} - p^2)A_1 = 0,$$

....

$$2mqA_m + [\tfrac{1}{4}(2m-1)^2 - p^2]A_{m-1} = 0;$$

therefore

$$v_p = \frac{e^{-nq}}{n^{\frac{1}{2}}} A_0 \left\{1 + \frac{\left(p^2 - \frac{1}{2^2}\right)}{2nq} + \frac{\left(p^2 - \frac{1}{2^2}\right)\left(p^2 - \frac{3^2}{2^2}\right)}{(2nq)^2 . 2} + \frac{\left(p^2 - \frac{1}{2^2}\right)\left(p^2 - \frac{3^2}{2^2}\right)\left(p^2 - \frac{5^2}{2^2}\right)}{(2nq)^3 . 3!} + \ldots\right\},$$

and A_0 alone remains to be determined; if we can determine A_0 for any value of p, we can find it for any other by means of equation (76). Now when $p = 0$,

$$v_0 = \int_0^\infty \frac{\cos nt \,.\, dt}{(q^2+t^2)^{\frac{1}{2}}}$$

and

$$\int_0^\infty \frac{\cos nt \,.\, dt}{(q^2+t^2)^{\frac{1}{2}}} = K(i \,.\, nq)$$

(Heine, *Kugelfunctionen*, vol. II. § 50), where K is the second kind of Bessel's function of zero order and $i = \sqrt{-1}$.

When nq is large,

$$K(i \, . \, nq) = \sqrt{(\tfrac{1}{2}\pi)} \frac{e^{-nq}}{(nq)^{\frac{1}{2}}}$$

(Heine, vol. I. § 61); hence

$$v_0 = \sqrt{(\tfrac{1}{2}\pi)} \frac{e^{-nq}}{(nq)^{\frac{1}{2}}} \left(1 - \frac{1}{2^2 \, . \, 2nq} + \frac{1 \, . \, 3^2}{2^5 \, . \, (2nq)^2} - \dots\right);$$

and, by equation (76), we find on comparing the coefficient of $\dfrac{e^{-nq}}{(nq)^{\frac{1}{2}}}$ that

$$A_0 = \sqrt{(\tfrac{1}{2}\pi)} \frac{q^{-p}}{1 \, . \, 3 \, . \, 5 \dots (2p-1)} \frac{e^{-nq}}{(nq)^{\frac{1}{2}}};$$

therefore

$$v_p = \sqrt{(\tfrac{1}{2}\pi)} \frac{q^{-p}}{1 \, . \, 3 \, . \, 5 \dots (2p-1)} \cdot \frac{e^{-nq}}{(nq)^{\frac{1}{2}}}$$

$$\times \left\{1 + \frac{\left(p^2 - \frac{1}{2^2}\right)}{2nq} + \frac{\left(p^2 - \frac{1}{2^2}\right)\left(p^2 - \frac{3^2}{2^2}\right)}{(2nq)^2 \, . \, 2} + \dots\right\},$$

$$\int_{-\infty}^{+\infty} \frac{\cos nt \, . \, dt}{(q^2+t^2)^{\frac{1}{2}(2p+1)}} = \sqrt{(2\pi)} \frac{n^p}{q^p} \frac{1}{1 \, . \, 3 \, . \, 5 \dots (2p-1)} \frac{e^{-nq}}{(nq)^{\frac{1}{2}}}$$

$$\times \left\{1 + \frac{\left(p^2 - \frac{1}{2^2}\right)}{2nq} + \frac{\left(p^2 - \frac{1}{2^2}\right)\left(p^2 - \frac{3^2}{2^2}\right)}{(2nq)^2 \, . \, 2} + \dots\right\} \dots\dots (77),$$

and this series converges rapidly when nq is large.

The other integrals in Q are of the form

$$\int_{-\infty}^{\infty} \frac{t \cos nt \, . \, dt}{(q^2+t^2)^{\frac{1}{2}(2p+1)}},$$

and these evidently vanish.

The integrals in P are of the forms

$$\int_{-\infty}^{\infty} \frac{\sin nt \, . \, dt}{(q^2+t^2)^{\frac{1}{2}(2p+1)}} \text{ and } \int_{-\infty}^{\infty} \frac{t \sin nt \, . \, dt}{(q^2+t^2)^{\frac{1}{2}(2p+1)}}.$$

The first of these evidently vanishes, and the second

$$= -\frac{d}{dn} \int_{-\infty}^{\infty} \frac{\cos nt \, . \, dt}{(q^2+t^2)^{\frac{1}{2}(2p+1)}},$$

and we have just found the value of the integral.

§ 34. We can now find the values of γ'_2 and α'_2.

By § 28,
$$\gamma'_2 = \frac{P\cos nt}{n} - \frac{Q\sin nt}{n},$$
where
$$P = \int_{-\infty}^{+\infty} \sin nt \,.\, \chi(t)\, dt,$$
$$Q = \int_{-\infty}^{+\infty} \cos nt \,.\, \chi(t)\, dt.$$

If we substitute for $\chi(t)$ its value, and evaluate the integrals by means of formula (77), and retain only the largest terms, we shall find

$$P = \frac{m\sqrt{(2\pi)}\,a^2b^2}{8k^9}\{4p^2\cos\epsilon\,(p\cos\epsilon - q)^2$$
$$- 4p\,(q^2-p^2)\,(q - p\cos\epsilon) + \cos\epsilon\,(q^2-p^2)^2\}\,.\,n^5\,\frac{e^{-nc/k}}{(nc/k)^{\frac{1}{2}}},$$
$$Q = \frac{m\sqrt{(2\pi)}\,a^2b^2}{8k^5}\sin\epsilon\,\{4p^2\,(p\cos\epsilon - q)^2 - (q^2-p^2)^2\}\,.\,n^5\,\frac{e^{-nc/k}}{(nc/k)^{\frac{1}{2}}}.$$

If the vortices move with equal velocities these expressions simplify very much and become

$$P = \frac{m\sqrt{(2\pi)}\,a^2b^2n^5}{8k^5}\cos\epsilon\,\frac{e^{-nc/k}}{(nc/k)^{\frac{1}{2}}},$$
$$Q = \frac{m\sqrt{(2\pi)}\,a^2b^2n^5}{8k^5}\sin\epsilon\,\frac{e^{-nc/k}}{(nc/k)^{\frac{1}{2}}},$$

so that
$$\gamma'_2 = \frac{m\sqrt{(2\pi)}\,a^2b^2n^4}{8k^5}\,\frac{e^{-nc/k}}{(nc/k)^{\frac{1}{2}}}\cos(nt+\epsilon) \;\ldots\ldots(78);$$

therefore
$$\alpha'_2 = -\frac{m\sqrt{(2\pi)}\,a^2b^2n^4}{\sqrt{3}\,.\,4k^5}\,\frac{e^{-nc/k}}{(nc/k)^{\frac{1}{2}}}\sin(nt+\epsilon) \;\ldots\ldots(79).$$

These equations represent twisted ellipses whose greatest ellipticity is

$$\frac{m\sqrt{(2\pi)}\,a^2bn^4}{\sqrt{3}\,.\,2k^5}\,\frac{e^{-nc/k}}{(nc/k)^{\frac{1}{2}}}.$$

The time of vibration is the corresponding free period.

§ 35. We can now sum up the effects of the collision of two vortices AB and CD.

The collisions must be divided into two classes, (1) those in which the shortest distance between the vortices is greater than twice the shortest distance between the directions of motion of the vortices; (2) those in which it is less.

Class I.

If the vortex CD be the first to intersect the shortest distance between the directions of motion of the vortices its radius is increased, and if its velocity is greater than the velocity of AB, resolved along the direction of motion of CD, it is bent towards the direction of motion of AB, and away from the plane containing the path of AB, and a parallel to that of CD. If its velocity is less than the value stated above it is bent from the direction of motion of AB and away from the plane containing the path of the centre of AB and a parallel to that of CD. This is the direction in which the path of CD is deflected if AB first intersects the shortest distance between the directions of motion of the vortices, but in this case the radius of CD is diminished.

Class II.

If the vortex CD be the first to intersect the shortest distance between the directions of motion of the vortices its radius is diminished by the collision. It is bent from or towards the direction of motion of AB according as its velocity is greater or less than the velocity of AB resolved along the direction of motion of CD, and away from or towards the plane containing the path of AB and a parallel to that of CD, according as the shortest distance between the vortices is greater or less than $\frac{2}{\sqrt{3}}$ times the shortest distance between their directions of motion. The deflection of AB with reference to this plane is the same whether AB or CD first intersect the shortest distance. If AB be the first to intersect the shortest distance, the radius of CD is increased, and the deflection of the path of CD relative to the direction of motion of AB is the opposite of that when CD was the first to intersect the shortest distance.

When the directions of motion of the vortices intersect, these results admit of much simpler statement, and, though included in Class I., it may be worth while to restate them. In this case the result is that the vortex which first passes through the point of intersection of the directions of motion of the vortices is deflected towards the direction of motion of the other; it increases in radius and energy, and its velocity is decreased; the other vortex is deflected in the same direction, it decreases in radius and energy, and its velocity is increased.

§ 36. Very closely allied to the problem of finding the action of two vortices on each other is the problem of finding the motion of one vortex when placed in a mass of fluid throughout which the distribution of velocity is known. We proceed to consider this problem, using the notation of § 14. Let Ω be the velocity potential of that part of the motion which is not due to the vortex ring itself. Let the equations to the central line of the vortex core be

$$\rho = a + \Sigma\,(\alpha_n \cos n\psi + \beta_n \sin n\psi),$$
$$z = \mathfrak{z} + \Sigma\,(\gamma_n \cos n\psi + \delta_n \sin n\psi).$$

Let $\pi\omega e^2$ be the strength of the vortex; let l, m, n be the direction-cosines of the normal to its plane, λ, μ, ν the direction-cosines of a radius vector of the vortex; then (§ 6)

$$\begin{aligned}
l &= \sin\theta\cos\epsilon,\\
m &= \sin\theta\sin\epsilon,\\
n &= \cos\theta,\\
\lambda &= \cos\epsilon\cos\theta\cos\psi - \sin\epsilon\sin\psi,\\
\mu &= \sin\epsilon\cos\theta\cos\psi + \cos\epsilon\sin\psi,\\
\nu &= -\sin\theta\cos\psi.
\end{aligned}$$

Let x, y, z be the coordinates of the centre of the vortex; if u, v, w be the velocities parallel to the axes of x, y, z at a point on the vortex ring, then, by Taylor's theorem,

$$u = \frac{d\Omega}{dx} + a\left(\lambda\frac{d}{dx} + \mu\frac{d}{dy} + \nu\frac{d}{dz}\right)\frac{d\Omega}{dx}$$
$$+ \tfrac{1}{2}a^2\left(\lambda\frac{d}{dx} + \mu\frac{d}{dy} + \nu\frac{d}{dz}\right)^2\frac{d\Omega}{dx} + \ldots,$$

with symmetrical expressions for v and w.

The velocity along the radius vector $= \lambda u + \mu v + \nu w$

$$= \left(\lambda\frac{d}{dx} + \mu\frac{d}{dy} + \nu\frac{d}{dz}\right)\Omega + a\left(\lambda\frac{d}{dx} + \mu\frac{d}{dy} + \nu\frac{d}{dz}\right)^2\Omega$$
$$+ \tfrac{1}{2}a^2\left(\lambda\frac{d}{dx} + \mu\frac{d}{dy} + \nu\frac{d}{dz}\right)^3\Omega + \ldots,$$

$\frac{da}{dt}$ = term in the expression for the velocity along the radius vector, which is independent of ψ.

As λ, μ, ν all involve ψ, the first powers of these quantities furnish nothing to this term.

$$\begin{aligned}
\lambda^2 &= \tfrac{1}{2}(1-l^2) + \tfrac{1}{2}\cos 2\psi(\cos^2\theta\cos^2\epsilon - \sin^2\epsilon) - \sin 2\psi\sin\epsilon\cos\epsilon\cos\theta,\\
\mu^2 &= \tfrac{1}{2}(1-m^2) + \tfrac{1}{2}\cos 2\psi(\cos^2\theta\sin^2\epsilon - \cos^2\epsilon) + \sin 2\psi\sin\epsilon\cos\epsilon\cos\theta,\\
\nu^2 &= \tfrac{1}{2}(1-n^2) + \tfrac{1}{2}\cos 2\psi\sin^2\theta,\\
\lambda\mu &= -\tfrac{1}{2}lm + \tfrac{1}{2}\cos 2\psi(1+\cos^2\theta)\sin\epsilon\cos\epsilon + \tfrac{1}{2}\sin 2\psi\cos\theta\cos 2\epsilon,\\
\lambda\nu &= -\tfrac{1}{2}ln + \tfrac{1}{2}\cos 2\psi(-\sin\theta\cos\theta\cos\epsilon) + \tfrac{1}{2}\sin 2\psi\sin\theta\sin\epsilon,\\
\mu\nu &= -\tfrac{1}{2}mn + \tfrac{1}{2}\cos 2\psi(-\sin\theta\cos\theta\sin\epsilon) - \tfrac{1}{2}\sin 2\psi\sin\theta\cos\epsilon.
\end{aligned}$$

The vortex itself contributes no term independent of ψ to the expression for the velocity along the radius vector; thus if the radius of the ring be small, we have approximately

$$\frac{da}{dt} = \frac{1}{2} a \left\{ (1 - l^2) \frac{d^2\Omega}{dx^2} + (1 - m^2) \frac{d^2\Omega}{dy^2} + (1 - n^2) \frac{d^2\Omega}{dz^2} - 2lm \frac{d^2\Omega}{dx\,dy} - 2ln \frac{d^2\Omega}{dx\,dz} - 2mn \frac{d^2\Omega}{dy\,dz} \right\};$$

or, since

$$\frac{d^2\Omega}{dx^2} + \frac{d^2\Omega}{dy^2} + \frac{d^2\Omega}{dz^2} = 0,$$

$$\frac{da}{dt} = -\frac{1}{2} a \left(l \frac{d}{dx} + m \frac{d}{dy} + n \frac{d}{dz} \right)^2 \Omega;$$

or, if $\frac{d}{dh}$ denote differentiation along the normal to the plane of the vortex ring,

$$\frac{da}{dt} = -\frac{1}{2} a \frac{d^2\Omega}{dh^2}.$$

From this equation we see that the radius of a vortex ring placed in a mass of fluid will increase or decrease according as the velocity along the normal to the plane of the vortex ring at the centre of the ring decreases or increases as we travel along a stream line through the centre. A simple application of this result is to the case when we have a fixed ring placed near a fixed barrier parallel to the plane of the ring. The effect of the barrier is to superpose on the distribution of velocity due to the vortex ring a velocity from the barrier which decreases as we recede from the barrier; it is this superposed velocity which affects the size of the ring, and, since the velocity decreases as we go along a stream line (which flows from the barrier), the preceding rule shews that the vortex will increase in size, which agrees with the well-known result for this case.

Let us now find how the vortex ring is deflected.

The velocity perpendicular to the plane of the vortex

$$= \frac{d\Omega}{dh} + a \left(\lambda \frac{d}{dx} + \mu \frac{d}{dy} + \nu \frac{d}{dz} \right) \frac{d\Omega}{dh} + \frac{1}{2} a^2 \left(\lambda \frac{d}{dx} + \mu \frac{d}{dy} + \nu \frac{d}{dz} \right)^2 \frac{d\Omega}{dh} + \ldots.$$

The coefficient of $\cos\psi$

$$= a \left(\cos\epsilon \cos\theta \frac{d}{dx} + \sin\epsilon \frac{d}{dy} - \sin\theta \frac{d}{dz} \right) \frac{d\Omega}{dh} + \text{terms in } a^3.$$

The coefficient of $\sin\psi$

$$= a \left(-\sin\epsilon \frac{d}{dx} + \sin\epsilon \cos\theta \frac{d}{dy} \right) \frac{d\Omega}{dh} + \text{terms in } a^3.$$

$\frac{d\gamma_1}{dt}$ = coefficient of $\cos\psi$ in the expression for the velocity perpendicular to the plane of the vortex.

The vortex itself contributes nothing to the coefficients of either $\cos\psi$ or $\sin\psi$ in the expression for the velocity perpendicular to the plane of the vortex (see equation 43).

Thus

$$\frac{d\gamma_1}{dt} = a\left(\cos\epsilon\cos\theta\frac{d}{dx} + \sin\epsilon\frac{d}{dy} - \sin\theta\frac{d}{dz}\right)\frac{d\Omega}{dh} \text{ approximately,}$$

$$\frac{d\delta_1}{dt} = a\left(-\sin\epsilon\frac{d}{dx} + \sin\epsilon\cos\theta\frac{d}{dy}\right)\frac{d\Omega}{dh}.$$

Now by § 6,

$$\frac{dl}{dt} = \frac{1}{a}\frac{d\delta_1}{dt}\sin\epsilon - \frac{1}{a}\frac{d\gamma_1}{dt}\cos\theta\cos\epsilon,$$

$$\frac{dm}{dt} = -\frac{1}{a}\frac{d\delta_1}{dt}\cos\epsilon - \frac{1}{a}\frac{d\gamma_1}{dt}\cos\theta\sin\epsilon,$$

$$\frac{dn}{dt} = \frac{1}{a}\frac{d\gamma_1}{dt}\sin\theta.$$

Substituting the values just found for $\frac{d\delta_1}{dt}$, $\frac{d\gamma_1}{dt}$ in these expressions, we find

$$\left.\begin{aligned} \frac{dl}{dt} &= l\,\frac{d^2\Omega}{dh^2} - \frac{d^2\Omega}{dh\,dx} \\ \frac{dm}{dt} &= m\,\frac{d^2\Omega}{dh^2} - \frac{d^2\Omega}{dh\,dy} \\ \frac{dn}{dt} &= n\,\frac{d^2\Omega}{dh^2} - \frac{d^2\Omega}{dh\,dz} \end{aligned}\right\} \qquad (80).$$

These equations enable us to find the orientation of the plane of the vortex at any time.

To find the change in the shape of the vortex, we have

$\frac{d\alpha_2}{dt}$ = coefficient of $\cos 2\psi$ in the expression for the velocity along the radius vector.

Now the vortex itself contributes to this coefficient the term

$$-\frac{2\omega e^2}{a^2}\log\frac{8a}{e}\cdot\gamma_2 \text{ (see equation 38).}$$

And if we pick out the coefficient of $\cos 2\psi$ arising from the velocity potential Ω, we shall find that it reduces to

$$-\tfrac{1}{2}a\left(\frac{d^2\Omega}{dh^2}+2\frac{d^2\Omega}{dk^2}\right),$$

where $\dfrac{d}{dk}$ denotes differentiation along an axis coinciding in direction with the radius of the vortex ring for which $\psi=\frac{1}{2}\pi$.

Thus $$\frac{d\alpha_2}{dt}=-\frac{2\omega e^2}{a^2}\log\frac{8a}{e}\,.\,\gamma_2-\tfrac{1}{2}a\left(\frac{d^2\Omega}{dh^2}+2\frac{d^2\Omega}{dk^2}\right).$$

Again,

$\dfrac{d\gamma_2}{dt}=$ coefficient of $\cos 2\psi$ in the expression for the velocity perpendicular to the plane of the vortex.

Now the vortex itself contributes to this coefficient the term

$$\tfrac{3}{2}\frac{\omega e^2}{a^2}\log\frac{8a}{e}\,.\,\alpha_2 \text{ (see equation 43).}$$

And if we pick out the coefficient of $\cos 2\psi$ arising from the velocity potential Ω, we shall find that it reduces to

$$-\tfrac{1}{4}a^2\left(\frac{d^2}{dh^2}+2\frac{d^2}{dk^2}\right)\frac{d\Omega}{dh}.$$

Thus

$$\frac{d\gamma_2}{dt}=\tfrac{3}{2}\frac{\omega e^2}{a^2}\log\frac{8a}{e}\,\alpha_2-\tfrac{1}{4}a^2\left(\frac{d^2}{dh^2}+2\frac{d^2}{dk^2}\right)\frac{d\Omega}{dh};$$

and this, with the preceding equation connecting $\dfrac{d\alpha_2}{dt}$ and γ_2, enables us to find α_2 and γ_2.

We have two exactly analogous equations connecting $d\beta_2/dt$ and δ_2, the only difference being that we substitute $\dfrac{d}{dk'}$ for $\dfrac{d}{dk}$, where $\dfrac{d}{dk'}$ denotes differentiation with respect to an axis passing through the centre and coinciding in direction with the radius of the vortex ring for which $\psi=0$.

§ 37. We can apply these equations to find the motion of a vortex ring which passes by a fixed obstacle. We shall suppose that the distance of the vortex from the obstacle is large compared with the diameter of the vortex, and that the obstacle is a sphere.

Let the plane containing the centre of the fixed sphere B, the centre of the vortex A, and a parallel to the direction of motion of the vortex be taken as the plane of xy. Let the axis of x be parallel to the direction of motion of the vortex. Let m' be the strength of the vortex, and a its radius.

The velocity potential due to the vortex at a point P

$$= \frac{1}{2} m' a^2 \frac{d}{dx} \cdot \left(\frac{1}{AP}\right) \text{ approximately.}$$

Now $$\frac{1}{AP} = \frac{1}{AB} + \frac{BP}{AB^2} Q_1 + \frac{BP^2}{AB^3} Q_2 + \dots \text{ (fig. 6)},$$

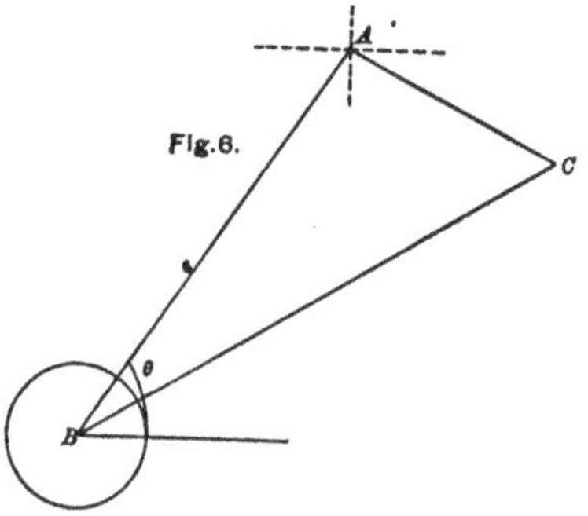

Fig. 6.

if $BP < AB$, and Q_1, Q_2...are spherical harmonics with AB for axis.

At the surface of the sphere the velocity parallel to x

$$= \frac{1}{2} m' a^2 \frac{d^2}{dx^2} \left(\frac{1}{AP}\right) = \frac{1}{2} m' a^2 \frac{3 \cos^2\theta - 1}{AB^3} + \text{smaller terms},$$

where θ is the angle AB makes with the axis of x.

The velocity parallel to the axis of y

$$= \frac{1}{2} m' a^2 \frac{d^2}{dx\, dy} \left(\frac{1}{AP}\right) = \frac{1}{2} m' a^2 \frac{3 \cos\theta \sin\theta}{AB^3} + \text{smaller terms.}$$

Now at the surface of the sphere the velocity must be entirely tangential, hence we must superpose a distribution of velocity, giving a radial velocity over the sphere equal and opposite to the radial velocity due to the vortex ring, *i. e.* equal to

$$-\frac{1}{2} \frac{m'}{AB^3} a^2 \left\{\frac{x}{b}(3\cos^2\theta - 1) + \frac{y}{b} 3 \cos\theta \sin\theta\right\},$$

if x and y be the coordinates of a point on the sphere, b the radius of the sphere. Let $AB = R$. Ω, the velocity potential which will give this radial velocity, is given by the equation

$$\Omega = -\frac{1}{4} \frac{m' b^3 a^2}{R^3} \left\{(3\cos^2\theta - 1) \frac{d}{dx} \frac{1}{r} + 3 \cos\theta \sin\theta \frac{d}{dy} \frac{1}{r}\right\},$$

where $r = BP$.

Ω is approximately the value of the velocity potential which produces the disturbance of the motion of the vortex.

The equation

$$\frac{da}{dt} = -\tfrac{1}{2}a\frac{d^2\Omega}{dh^2}$$

becomes in this case

$$\frac{da}{dt} = \tfrac{1}{8}\frac{m'a^3b^3}{R^3}\left\{(3\cos^2\theta - 1)\frac{d^3}{dx^3}\frac{1}{r} + 3\cos\theta\sin\theta\frac{d^3}{dx^2dy}\frac{1}{r}\right\}.$$

Now
$$\frac{d^3}{dx^3}\left(\frac{1}{r}\right) = -\frac{3(5\cos^3\theta - 3\cos\theta)}{R^4},$$

$$\frac{d^3}{dx^2dy}\left(\frac{1}{r}\right) = \frac{3\sin\theta(1 - 5\cos^2\theta)}{R^4}.$$

We must express the quantities on the right-hand side of the equation in terms of the time.

Let us measure the time from the instant when the line joining the centre of the sphere to the centre of the vortex is perpendicular to the direction of motion of the vortex. Let u be the velocity of the vortex; then we have, accurately if the motion were undisturbed, and very approximately as the motion of the vortex is only slightly disturbed,

$$R^2 = c^2 + u^2t^2,$$

$$\cos\theta = \frac{ut}{(c^2 + u^2t^2)^{\frac{1}{2}}},$$

$$\sin\theta = \frac{c}{(c^2 + u^2t^2)^{\frac{1}{2}}},$$

where c is the shortest distance between the centre of the vortex and the centre of the sphere.

Substituting we find

$$\frac{da}{dt} = -\tfrac{3}{2}\frac{u^3t^3m'a^3b^3}{(c^2 + u^2t^2)^5};$$

thus the vortex expands until it gets to its shortest distance from the centre of the sphere, after passing its shortest distance it ceases to expand and begins to contract.

Integrating the differential equation, we get

$$a = a_0\left\{1 + \frac{1}{16u}\frac{m'a_0^2b^3}{R^6}(1 + 3\cos^2\theta)\right\},$$

where a_0 is the value of a before the vortex got near the sphere.

Thus we see that the radius is the same after the vortex has passed quite away from the sphere as it was before it got near to it, since in both cases $R = \infty$; in intermediate positions it is always greater.

The greatest value of the radius is

$$a_0\left(1+\frac{1}{16u}\frac{m'a_0^2b^3}{c^6}\right);$$

the greatest increase in the radius is thus proportional to the volume of the sphere, and inversely proportional to the sixth power of the shortest distance between the vortex and the sphere.

§ 38. To find the way in which the direction of motion of the vortex is altered we have, if l, m are the x and y direction cosines of the normal to its plane,

$$\frac{dm}{dt}=m\frac{d^2\Omega}{dx^2}-\frac{d^2\Omega}{dxdy}.$$

Now in the undisturbed motion $m=0$, so we may write this equation

$$\frac{dm}{dt}=-\frac{d^2\Omega}{dxdy},$$

$$\frac{dm}{dt}=\tfrac{1}{4}\frac{m'b^3a^2}{R^3}\left\{(3\cos^2\theta-1)\frac{d^3}{dx^2dy}\left(\frac{1}{r}\right)+3\cos\theta\sin\theta\frac{d^3}{dy^2dx}\left(\frac{1}{r}\right)\right\}.$$

Now
$$\frac{d^3}{dx^2dy}\left(\frac{1}{r}\right)=\frac{3y(r^2-5x^2)}{r^7},$$

$$\frac{d^3}{dy^2dx}\left(\frac{1}{r}\right)=\frac{3x(r^2-5y^2)}{r^7}.$$

Substituting these values, we find

$$\frac{dm}{dt}=-\tfrac{3}{4}\frac{m'b^3a^2}{R^7}\sin\theta\,(1+4\cos^2\theta);$$

thus $\frac{dm}{dt}$ is always negative, or the vortex moves as if attracted by the sphere; expressing the right-hand side in terms of the time, we get

$$\frac{dm}{dt}=-\tfrac{3}{4}m'b^3a^2c\left\{\frac{5}{(c^2+u^2t^2)^4}-\frac{4c^2}{(c^2+u^2t^2)^5}\right\}.$$

Integrating both sides from $t=-\infty$ to $t=+\infty$, we find that m, the whole angle turned through by the vortex, is given by the equation

$$m=-\tfrac{45}{128}\frac{\pi m'b^3a^2}{c^6},$$

and this effect varies inversely as the sixth power of the shortest distance between the vortex ring and the sphere, and directly as the volume of the sphere. Sir William Thomson shewed by general reasoning that a vortex passing near a fixed solid will appear to be attracted by it ("Vortex Motion," *Edinburgh Transactions*, vol. XXV. p. 229); and this result agrees with the results we have obtained for the sphere.

PART III.

Linked Vortices.

§ 39. We must now pass on to discuss the case of Linked Vortices. We shall suppose that we have two vortex rings linked one through the other in such a way that the shortest distance between the vortex rings at any point is small compared with the radius of the aperture of either vortex ring, but large compared with the radius of the cross section of either of them. Thus, the circumstances in this case are the opposite to those in the case we have just been considering, when the shortest distance between the vortices was large compared with the diameter of either.

In the present case it is important to examine the changes in the shape of the cross section of the vortices, in order to see that they remain approximately circular. We shall, therefore, discuss this problem first.

Since the distance between the vortices is very small compared with the radii of the apertures of the vortices, the changes in their cross sections will be very approximately the same as the changes in the cross sections of two infinitely long straight cylindrical vortex columns placed in the same mass of fluid in such a manner that the distance between them is great compared with the radius of either of their cross sections.

We shall prove that if the cross sections of two such vortex columns are at any moment approximately circular they will always remain so.

We must first find the velocity potential due to such a vortex column.

Let the equation to the cross section be

$$\rho = a + \alpha_n \cos n\theta + \beta_n \sin n\theta,$$

where α_n and β_n are small compared with a, the mean radius of the section. Let ω be the angular velocity of molecular rotation.

The stream function ψ due to this distribution of vorticity is given by the equation

$$\psi = -\frac{1}{\pi}\iint \omega \log r \, dx' \, dy'$$

(Lamb's *Treatise on the Motion of Fluids*, § 138, equation 33), where r is the distance of the points x, y from the points x', y'.

Thus ψ is the potential of matter of density $-\frac{\omega}{2\pi}$ distributed over the cross section.

At a point outside the cylinder let

$$\psi = C - \omega a^2 \log r + (A_n \cos n\theta + B_n \sin n\theta)\frac{a^n}{r^n} \dots (81).$$

At a point inside the cylinder let

$$\psi = C' - \tfrac{1}{2}\omega r^2 + (A'_n \cos n\theta + B'_n \sin n\theta)\frac{r^n}{a^n} \dots\dots (82).$$

Thus, since ψ is continuous, these two values must be equal at the surface of the cylinder; thus, if we substitute

$$r = a + \alpha_n \cos n\theta + \beta_n \sin n\theta,$$

we may equate the coefficients of $\cos n\theta$ and $\sin n\theta$ in the two expressions for ψ.

Doing this we get, neglecting powers higher than the first of α_n and β_n,

$$-\omega a \alpha_n + A_n = -\omega a \alpha_n + A'_n,$$
$$-\omega a \beta_n + B_n = -\omega a \beta_n + B'_n;$$

or

$$A_n = A'_n,$$
$$B_n = B'_n.$$

The differential coefficients of ψ are continuous; thus the two values of $\frac{d\psi}{dr}$ must be the same at the surface of the cylinder; differentiating both expressions for ψ with respect to r, putting

$$r = a + \alpha_n \cos n\theta + \beta_n \sin n\theta,$$

and equating the coefficients of $\cos n\theta$ and $\sin n\theta$, we find

$$\omega \alpha_n - \frac{nA_n}{a} = \frac{nA'_n}{a} - \omega \alpha_n,$$

$$\omega \beta_n - \frac{nB_n}{a} = \frac{nB'_n}{a} - \omega \beta_n.$$

Solving these equations, we find

$$A_n = \frac{a\omega\alpha_n}{n}, \quad B_n = \frac{a\omega\beta_n}{n}.$$

Thus at a point outside the cylinder,

$$\psi = C - \omega a^2 \log r + \frac{a\omega}{n}(\alpha_n \cos n\theta + \beta_n \sin n\theta)\frac{a^n}{r^n} \quad \ldots(83).$$

We can now find the time of vibration of a single vortex column whose section differs slightly from the circular form.

For if $\rho = a + \alpha_n \cos n\theta + \beta_n \sin n\theta$ be the equation to the cross section, then, since the surface always consists of the same particles of the fluid, using the theorem that if $F(x, y, z, t) = 0$ be the equation to such a surface,

$$\frac{dF}{dt} + u\frac{dF}{dx} + v\frac{dF}{dy} + w\frac{dF}{dz} = 0,$$

we get

$$\mathfrak{R} = \frac{d\alpha_n}{dt}\cos n\theta + \frac{d\beta_n}{dt}\sin n\theta - n(\alpha_n \sin n\theta - \beta_n \cos n\theta)\,\Theta \quad \ldots(84),$$

where $\mathfrak{R}$ is the velocity of the fluid at the surface of the cylinder along the radius vector and Θ its angular velocity round the axis of the cylinder.

Now

$$\mathfrak{R} = \frac{d\psi}{rd\theta},$$

$$\Theta = -\frac{1}{r}\frac{d\psi}{dr}.$$

Thus, when $r = a + \alpha_n \cos n\theta + \beta_n \sin n\theta$,

$$\left.\begin{aligned} \mathfrak{R} &= -\omega(\alpha_n \sin n\theta - \beta_n \cos n\theta) \\ \Theta &= \omega - \frac{\omega}{a}(\alpha_n \cos n\theta + \beta_n \sin n\theta) \end{aligned}\right\} \quad \ldots\ldots\ldots\ldots(85),$$

neglecting squares of α_n and β_n.

Hence substituting in equation (84) and neglecting all powers of α_n and β_n above the first, we get

$$-\omega(\alpha_n \sin n\theta - \beta_n \cos n\theta) = \frac{d\alpha_n}{dt}\cos n\theta + \frac{d\beta_n}{dt}\sin n\theta - n\omega(\alpha_n \sin n\theta - \beta_n \cos n\theta):$$

equating coefficients of $\cos n\theta$ and $\sin n\theta$, we get

$$\frac{d\alpha_n}{dt} = -(n-1)\,\omega\beta_n,$$

$$\frac{d\beta_n}{dt} = (n-1)\,\omega\alpha_n;$$

therefore

$$\frac{d^2\alpha_n}{dt^2} + (n-1)^2\omega^2\alpha_n = 0,$$

or

$$\alpha_n = A \cos \{(n-1)\omega t + \beta\},$$
$$\beta_n = A \sin \{(n-1)\omega t + \beta\},$$

where A and β are arbitrary constants.

Thus $$r = a + A \cos [\{n\theta - (n-1)\omega t\} - \beta] \quad \text{............(86).}$$

Thus the section never differs much from a circle, and the disturbance in the shape travels round the cylinder in the time

$$\frac{2\pi}{(n-1)\omega}.$$

These results agreed with those stated by Sir William Thomson in his paper on "Vortex Atoms" (*Phil. Mag.* 1867), and proved in his paper "On the Vibration of a Columnar Vortex." *Proceedings of the Royal Society of Edinburgh*, March 1, 1880; reprinted in *Phil. Mag.*, Sep. 1880.

§ 40. Let us now consider the case when there are two vortex columns in the fluid (fig. 7).

Fig. 7.

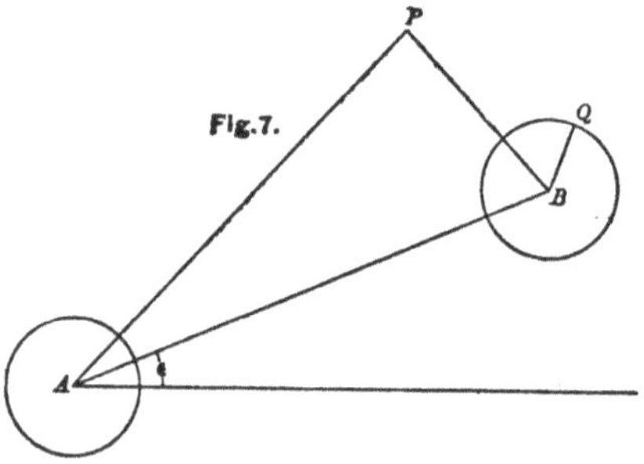

Let $$\rho = a + \Sigma (\alpha_n \cos n\theta + \beta_n \sin n\theta)$$

be the equation to the cross section of the one with A as centre, and let

$$\rho' = b + \Sigma (\alpha_n' \cos n\theta' + \beta'_n \sin n\theta')$$

be the equation to the cross section of the one with B as centre, ρ being measured from A and ρ' from B.

Let c be the distance AB between their centres, and ϵ the angle AB makes with the initial line.

Then the stream function ψ due to the two vortex columns at a point P is given by the equation

$$\psi = C - \omega a^2 \log r + \Sigma \frac{a\omega}{n} (\alpha_n \cos n\theta + \beta_n \sin n\theta) \frac{a^n}{r^n}$$
$$- \omega' b^2 \log r' + \Sigma \frac{b\omega'}{n} (\alpha'_n \cos n\theta' + \beta'_n \sin n\theta') \frac{b^n}{r'^n},$$

where $r = AP$, $r' = BP$, and θ, θ' are the angles AP and BP make with the initial line, ω and ω' are the angular velocities of molecular rotation of the two vortex columns.

We shall want to use the current function at the surface of both the cylinders, thus it will be convenient to find a method of transforming that part of the stream function where the coordinates used are measured from A as origin to coordinates with B as origin, and *vice versâ*. To do this we shall use the following lemma, which may be easily proved by trigonometry.

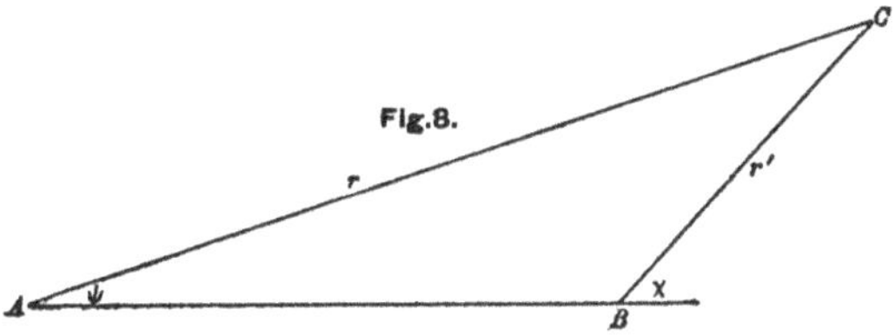

Fig. 8.

Lemma.

§ 41. If $AP = r$, $BP = r'$, $< PAB = \psi$, $< PBC = \chi$, $AB = c$. Then if $r' < c$

$$\frac{\sin n\psi}{r^n} = \frac{n}{c^n}\left\{\left(\frac{r'}{c}\right)\sin\chi - \frac{n+1}{1.2}\left(\frac{r'}{c}\right)^2\sin 2\chi + \frac{n+1.n+2}{1.2.3}\left(\frac{r'}{c}\right)^3\sin 3\chi + \ldots\right\},$$

$$\frac{\cos n\psi}{r^n} = \frac{1}{c^n}\left\{1 - n.\frac{r'}{c}\cos\chi + \frac{n.n+1}{1.2}\left(\frac{r'}{c}\right)^2\cos 2\chi - \frac{n.n+1.n+2}{1.2.3}\left(\frac{r'}{c}\right)^3\cos 3\chi + \ldots\right\},$$

if $r' < c$,

$$\frac{\sin n\psi}{r^n} = \frac{1}{r'^n}\left\{\sin n\chi - n\frac{c}{r'}\sin(n+1)\chi + \frac{n.n+1}{1.2}\left(\frac{c}{r'}\right)^2\sin(n+2)\chi\ldots\right\},$$

$$\frac{\cos n\psi}{r^n} = \frac{1}{r'^n}\left\{\cos n\chi - n\frac{c}{r'}\cos(n+1)\chi + \frac{n.n+1}{1.2}\left(\frac{c}{r'}\right)^2\cos(n+2)\chi\ldots\right\}.$$

§ 42. Again
$$\log r = \tfrac{1}{2}\log r^2 = \tfrac{1}{2}\log(r'^2 + c^2 + 2cr'\cos\chi).$$

If $c > r'$,

$$\log r = \log c + \frac{r'}{c}\cos\chi - \tfrac{1}{2}\frac{r'^2}{c^2}\cos 2\chi + \tfrac{1}{3}\frac{r'^3}{c^3}\cos 3\chi\ldots.$$

If $c < r'$,

$$\log r = \log r' + \frac{c}{r}\cos\chi - \frac{1}{2}\frac{c^2}{r'^2}\cos 2\chi + \frac{1}{3}\frac{c^3}{r'^3}\cos 3\chi \ldots.$$

We can now find the effect of the vortex columns on each other.

For if $\mathfrak{R}$ be the radial velocity of a point Q on one of the vortex columns relative to B the centre of that vortex column, and $b\Theta$ the velocity of Q relative to B, perpendicular to BQ, then as before

$$\mathfrak{R} = \frac{d\alpha'_n}{dt}\cos n\theta + \frac{d\beta'_n}{dt}\sin n\theta' - n(\alpha'_n \sin n\theta - \beta'_n \cos n\theta)\Theta \ldots (87).$$

Now, the part of $\mathfrak{R}$ due to the vortex column with B as centre

$$= -\omega'(\alpha'_n \cos n\theta' - \beta'_n \sin n\theta'),$$

the part of $\mathfrak{R}$ due to the term $-\omega a^2 \log r$ in the stream function

$$= -\omega a^2 \left\{\frac{b}{c^2}\sin 2(\theta' - \epsilon) - \frac{b^2}{c^3}\sin 3(\theta' - \epsilon) + \frac{b^3}{c^4}\sin 4(\theta' - \epsilon) \ldots\right\},$$

the term $\qquad \dfrac{a\omega}{n}(\alpha_n \cos n\theta + \beta_n \sin n\theta)\dfrac{a^n}{r^n}$,

$$\text{gives } -a\omega\frac{a^n}{c^{n+1}}(\alpha_n \cos n\epsilon + \beta_n \sin n\epsilon)(n+1)$$
$$\times\left\{\frac{b}{c}\sin 2(\theta' - \epsilon) - \frac{n+2}{1.2}\left(\frac{b}{c}\right)^2 \sin 3(\theta' - \epsilon) + \ldots\right\}$$
$$-a\omega\frac{a^n}{c^{n+1}}(\beta_n \cos n\epsilon - \alpha_n \sin n\epsilon)(n+1)$$
$$\times\left\{\frac{b}{c}\cos 2(\theta' - \epsilon) - \frac{(n+2)}{1.2}\left(\frac{b}{c}\right)^2 \cos 3(\theta' - \epsilon) + \ldots\right\}.$$

Since α_n, β_n, and $\dfrac{a}{c}$ are all small quantities, as we are neglecting the squares of small quantities, we may neglect these terms which involve quantities of the order of α^2_n; and for the same reason, we may in equation (87) put $\Theta = \omega'$, since it only differs from it by small quantities of the order α_n and $\dfrac{a}{c}$, and in that equation Θ is multiplied by quantities of this order.

Substituting these values for $\mathfrak{R}$ and Θ in equation (87), and equating the coefficients of $\cos\theta'$, and $\sin\theta'$ on each side of the equations, we get

$$\frac{d\alpha'_1}{dt} = 0,\ \frac{d\beta'_1}{dt} = 0,$$

or, as α'_1 and β'_1 are zero initially we get $\alpha'_1 = 0$, $\beta'_1 = 0$, and similarly $\alpha_1 = 0, \beta_1 = 0$; and thus the motion of the centre of gravity of either vortex column is not disturbed. If we equate the coefficients of $\cos 2\theta'$ and $\sin 2\theta'$ on each side of equation (87), we get

$$\frac{d\alpha'_2}{dt} + \omega'\beta'_2 = \frac{\omega a^2 b \sin 2\epsilon}{c^2},$$

and

$$\frac{d\beta'_2}{dt} - \omega'\alpha'_2 = -\frac{\omega a^2 b \cos 2\epsilon}{c^2}.$$

Now AB travels round approximately uniformly with an angular velocity n, where $n = \frac{\omega a^2 + \omega' b^2}{c^2}$, this value of n follows at once if we remember that the centre of gravity of the two vortex columns remains at rest.

Thus taking the initial position of AB as the initial line from which to measure our angles, we have $\epsilon = nt$.

Thus

$$\frac{d\alpha'_2}{dt} + \omega'\beta'_2 = \frac{\omega a^2 b}{c^2} \sin 2nt,$$

$$\frac{d\beta'_2}{dt} - \omega'\alpha'_2 = -\frac{\omega a^2 b}{c^2} \cos 2nt;$$

therefore

$$\frac{d^2\alpha'_2}{dt^2} + \omega'^2\alpha'_2 = \frac{\omega a^2 b}{c^2}(2n + \omega') \cos 2nt;$$

therefore

$$\alpha'_2 = A \cos(\omega' t + \beta) + \frac{\omega a^2 b}{c^2} \frac{(2n + \omega') \cos 2nt}{\omega'^2 - 4n^2} \quad \text{......(88).}$$

Now, let $\alpha_2, \beta_2 = 0$ initially, then $d\alpha'_2/dt = 0$ initially, and we get

$$\left.\begin{aligned} \alpha'_2 &= \frac{\omega a^2 b}{c^2(\omega' - 2n)}(\cos 2nt - \cos \omega' t) \\ \beta'_2 &= \frac{\omega a^2 b}{c^2(\omega' - 2n)}(\sin 2nt - \sin \omega' t) \end{aligned}\right\} \quad \text{............(89).}$$

Thus the cross section at any instant is an ellipse. This ellipse does not, however, remain of the same shape, but vibrates about the circular form; the maximum ellipticity is proportional to $\frac{\omega a^2 b}{c^2(\omega' - 2n)}$, and thus varies inversely as the square of the distance between the vortex columns. The vibration has two periods, a long one $\frac{\pi}{n}$ and a short one $\frac{2\pi}{\omega'}$.

The terms in α_3, β_3 will involve $\frac{a^3}{c^3}$, and thus will be relatively unimportant, as α_2, β_2 only involve the square of $\frac{a}{c}$; the same reasoning applies *à fortiori* to α_n and β_n when n is greater than three.

§ 42. Our investigation of the motion of two infinite cylindrical vortices shews that to retain an approximately circular cross section the vortices must be at a distance from each other large compared with the diameter of the cross section of either. If we consider a portion of two linked vortices near each other, and regard them as straight, which we may do if the distance between them is small compared with the radius of the aperture of either, we see that the vortices will spin round each other with an angular velocity $\frac{m+m'}{\pi d^2}$ when m and m' are the strength of the two vortices, and d the shortest distance between the two parts of the vortices we are considering; thus, if the motion is to be steady, we must have this angular velocity approximately constant all round the vortices, and therefore d^2 must be approximately constant all round the vortices.

To get a clear conception of the way the vortices, supposed for the moment of equal strength, are linked, we may regard them as linked round an anchor ring whose transverse section is small compared with its aperture, the manner of linking being such that there are always portions of the two vortices at opposite extremities of a diameter of a transverse section of the anchor ring. The shortest distance between pieces of the two vortices is then approximately constant, and equal to the diameter of the transverse section of the anchor ring.

Let us suppose that the vortex is linked r times round the anchor ring, then the equation to the central line of vortex core may be written

$$\rho = a + \alpha_1 \cos\theta + \beta_1 \sin\theta + \ldots \alpha_r \cos r\theta + \beta_r \sin r\theta$$
$$+ \ldots \alpha_n \cos n\theta + \beta_n \sin n\theta + \ldots$$
$$z = \mathfrak{z} + \gamma_1 \cos\theta + \delta_1 \sin\theta + \ldots \gamma_r \cos r\theta + \delta_r \sin r\theta$$
$$+ \ldots \gamma_n \cos n\theta + \delta_n \sin n\theta + \ldots.$$

Let the equations to the second vortex differ from these only in having accents affixed to the letters. Here α_1, β_1; γ_1, δ_1; α'_1, β'_1; γ'_1, δ'_1, &c. are all small in comparison with a and a', but α_r, β_r; γ_r, δ_r; α'_r, β'_r; γ'_r, δ'_r are large compared with the others, so that in the expression for the velocities due to the vortex rings we shall go to the squares of these quantities, but only retain the first powers of the other quantities denoted by the Greek letters. Let m be the strength of the vortex whose equation was first written, which we shall call vortex (I), m' the strength of the other, which we shall call vortex (II). Let e and e' be the radii of the cross sections of vortices (I) and (II) respectively.

Let ${}_{11}A_n$ denote the value of the quantity we denoted in § 13 by A_n, due to the vortex (I) at a point on the surface of itself.

${}_{12}A_n$ the value of the quantity A_n due to the vortex (I) at a point on the surface of vortex (II).

${}_{21}A_n$ the value of the quantity A_n due to the vortex (II) at a point on the surface of the vortex (I).

${}_{22}A_n$ the value of the quantity A_n due to the vortex (II) at a point on the surface of itself.

Now, from equations (11) and (14) the terms of the first order in α_n, &c., in the expression for the velocity along the radius vector due to the vortex (I) at the surface of the vortex (II) are

$$\tfrac{1}{2} ma \Sigma \{{}_{12}A_1 (\gamma'_n \cos n\psi + \delta'_n \sin n\psi) \\ + \tfrac{1}{2} [(n-1)\, {}_{12}A_{n+1} - (n+1)\, {}_{12}A_{n-1}] (\gamma_n \cos n\psi + \delta_n \sin n\psi)\}.$$

If we suppose the two vortices wound round an anchor ring, of diameter d, in such a way that there are always portions of the two vortices at opposite extremities of a diameter of the transverse section, then in the expression for A_n given in equation (35) we must put $x = \dfrac{d^2}{2a^2}$. Substituting this value of A_n and retaining only the most important terms, we find that the velocity along the radius vector of the vortex (II) due to the vortex (I)

$$= \frac{m}{4\pi a^2} \Sigma \left\{ (\gamma'_n \cos n\psi + \delta'_n \sin n\psi) \left(\frac{4a^2}{d^2} - \tfrac{3}{4} \log \frac{64a^2}{d^2} \right) \right. \\ \left. + (\gamma_n \cos n\psi + \delta_n \sin n\psi) \left(-\frac{4a^2}{d^2} - (n^2 - \tfrac{3}{4}) \log \frac{64a^2}{d^2} \right) \right\}.$$

By equation (38) we see that the velocity along the radius vector of the vortex (II) due to this vortex itself

$$= -\frac{m'}{4\pi a'^2} \Sigma (\gamma'_n \cos n\psi + \delta'_n \sin n\psi)\, n^2 \log \frac{64a'^2}{e'^2}.$$

But from the equation

$$\rho' = a' + \Sigma (\alpha'_n \cos n\psi + \beta'_n \sin n\psi) + e' \cos \phi,$$

we see that if we only retain the first powers of the quantities α'_n, β'_n, the velocity along the radius vector

$$= \frac{d\alpha'_n}{dt} \cos n\psi + \frac{d\beta'_n}{dt} \sin n\psi,$$

equating the coefficients of $\cos n\psi$ and $\sin n\psi$ in this expression for the velocity and in the expression just found, we find

$$\frac{d\alpha'_n}{dt}=\frac{m}{4\pi a^2}\left\{\gamma'_n\left(\frac{4a^2}{d^2}-\tfrac{3}{4}\log\frac{64a^2}{d^2}\right)-\gamma_n\left(\frac{4a^2}{d^2}+(n^2-\tfrac{3}{4})\log\frac{64a^2}{d^2}\right)\right\}$$
$$-\frac{m'}{4\pi a'^2}\gamma_n'\,n^2\log\frac{64a'^2}{e'^2},$$

$$\frac{d\beta'_n}{dt}=\frac{m}{4\pi a^2}\left\{\delta'_n\left(\frac{4a^2}{d^2}-\tfrac{3}{4}\log\frac{64a^2}{d^2}\right)-\delta_n\left(\frac{4a^2}{d^2}+(n^2-\tfrac{3}{4})\log\frac{64a^2}{d^2}\right)\right\}$$
$$-\frac{m'}{4\pi a'^2}\delta'_n\,n^2\log\frac{64a'^2}{e'^2}.$$

From equations (16) and (17) the terms of the first order in α_n, &c., in the expression for the velocity perpendicular to the plane of vortex (II) due to vortex (I)

$$=\tfrac{1}{2}ma\,(2a_{12}A_0-a'_{12}A_1)-\tfrac{1}{2}ma\,(\alpha'_n\cos n\psi+\beta'_n\sin n\psi)_{12}A_1$$
$$+\tfrac{1}{2}ma^2(\alpha'_n\cos n\psi+\beta'_n\sin n\psi)\,\tfrac{1}{2}\left(\frac{d}{dr'}+\frac{d}{dR}\right)(2_{12}A_0-{}_{12}A_1)$$
$$+\tfrac{1}{2}m\,(\alpha_n\cos n\psi+\beta_n\sin n\psi)\left\{\tfrac{1}{2}a^2\left(\frac{d}{dr'}+\frac{d}{dR}\right)\{{}_{12}A_n-\tfrac{1}{2}({}_{12}A_{n+1}+{}_{12}A_{n-1})\}\right.$$
$$\left.+2a_{12}A_n+\tfrac{1}{2}a\,\{(n-1)\,{}_{12}A_{n-1}-(n+1)\,{}_{12}A_{n+1}\}\right\},$$

where, before differentiation, the A's are to be regarded as functions of r' and R, and after differentiation we put

$$r'=a+\alpha_r\cos r\psi+\beta_r\sin r\psi,$$
$$R=a'+\alpha'_r\cos r\psi+\beta'_r\sin r\psi,$$

and retain the largest terms; the quantities α_r, β_r, α'_r, β'_r, have each $\tfrac{1}{2}d$ for their maximum value. If we substitute in these expressions the values for the quantities denoted by the A's given in equation (35), and put $x=\dfrac{d^2}{2a^2}$, we find that the velocity perpendicular to the plane of vortex (II) due to vortex (I)

$$=\frac{m}{2\pi a}\left(\log\frac{8a}{d}-1\right)-\frac{m}{4\pi a^2}(\alpha'_n\cos n\psi+\beta'_n\sin n\psi)\left(\frac{4a^2}{d^2}+\tfrac{3}{4}\log\frac{64a^2}{a^2}\right)$$
$$+\frac{m}{4\pi a^2}(\alpha_n\cos n\psi+\beta_n\sin n\psi)\left(\frac{4a^2}{d^2}+(n^2-\tfrac{1}{4})\log\frac{64a^2}{d^2}\right),$$

if we go to the first powers only of the quantities denoted by the Greek letters.

The velocity perpendicular to the plane of the vortex (II) due to this vortex itself, is by equation (43)

$$\frac{m'}{2\pi a'}\left(\log\frac{8a'}{e'}-1\right)+\frac{m'}{4\pi a'^2}(n^2-1)\log\frac{64a'^2}{e'^2}\,(\alpha_n'\cos n\psi+\beta_n'\sin n\psi).$$

But from the equation

$$z = \mathfrak{z} + \Sigma\,(\gamma'_n \cos n\psi + \delta'_n \sin n\psi),$$

we see, as in equation (40), that the velocity perpendicular to the plane of the vortex, is

$$\frac{d\mathfrak{z}}{dt} + \Sigma\left(\frac{d\gamma_n}{dt}\cos n\psi + \frac{d\delta_n}{dt}\sin n\psi\right).$$

Hence, equating the constant terms and the coefficients of $\cos n\psi$ and $\sin n\psi$, in this expression, and the expression we have just found for the same quantity, we get

$$\frac{d\mathfrak{z}}{dt} = \frac{m}{2\pi a}\left(\log\frac{8a}{d} - 1\right) + \frac{m'}{2\pi a'}\left(\log\frac{8a'}{e'} - 1\right)$$

$$\frac{d\gamma'_n}{dt} = \frac{m}{4\pi a^2}\left\{\alpha_n\left(\frac{4a^2}{d^2} + (n^2 - \tfrac{1}{4})\log\frac{64a^2}{d^2}\right) - \alpha'_n\left(\frac{4a^2}{d^2} + \tfrac{3}{4}\log\frac{64a^2}{d^2}\right)\right\}$$
$$+ \frac{m'}{4\pi a'^2}\,\alpha'_n\,(n^2 - 1)\log\frac{64a'^2}{e'^2},$$

$$\frac{d\delta'_n}{dt} = \frac{m}{4\pi a^2}\left\{\beta_n\left(\frac{4a^2}{d^2} + (n^2 - \tfrac{1}{4})\log\frac{64a^2}{d^2}\right) - \beta'_n\left(\frac{4a^2}{d^2} + \tfrac{3}{4}\log\frac{64a^2}{d^2}\right)\right\}$$
$$+ \frac{m'}{4\pi a'^2}\,\beta'_n\,(n^2 - 1)\log\frac{64a'^2}{e'^2}.$$

In the case we are considering the mean radii of the vortices are equal, thus $a = a'$.

If we write for the sake of brevity,

$$\left.\begin{aligned}
L &= \frac{4a^2}{d^2} - \tfrac{3}{4}\log\frac{64a^2}{d^2}\\
M' &= n^2\log\frac{64a^2}{e'^2}\\
N &= \frac{4a^2}{d^2} + (n^2 - \tfrac{3}{4})\log\frac{64a^2}{d^2}\\
P &= \frac{4a^2}{d^2} + \tfrac{3}{4}\log\frac{64a^2}{d^2}\\
Q' &= (n^2 - 1)\log\frac{64a^2}{e'^2}\\
R &= \frac{4a^2}{d^2} + (n^2 - \tfrac{1}{4})\log\frac{64a^2}{d^2}
\end{aligned}\right\}\quad\ldots\ldots\ldots\ldots(90).$$

Then our equations become

$$\frac{d\alpha'_n}{dt} = \frac{1}{4\pi a^2}\{\gamma'_n(mL - m'M') - \gamma_n mN\}\ldots\ldots\ldots\ldots(91),$$

$$\frac{d\gamma'_n}{dt} = \frac{1}{4\pi a^2}\{\alpha'_n(m'Q' - mP) + \alpha_n mR\} \quad \text{.............(92).}$$

If we go to the vortex (I), we get

$$\frac{d\mathfrak{z}}{dt} = \frac{m'}{2\pi a}\log\frac{8a}{d} + \frac{m}{2\pi a}\log\frac{8a}{e} - \frac{m+m'}{2\pi a},$$

$$\frac{d\alpha_n}{dt} = \frac{1}{4\pi a^2}\{\gamma_n(m'L - mM) - \gamma'_n m'N\},$$

$$\frac{d\gamma_n}{dt} = \frac{1}{4\pi a^2}\{\alpha_n(mQ - m'P) + \alpha'_n m'R\},$$

where M and Q are what M' and Q' become when e is written for e'.

Equating the two values of $\frac{d\mathfrak{z}}{dt}$, we must have

$$\frac{m}{2\pi a}\log\frac{8a}{d} + \frac{m'}{2\pi a}\log\frac{8a}{e'} = \frac{m'}{2\pi a}\log\frac{8a}{d} + \frac{m}{2\pi a}\log\frac{8a}{e},$$

or

$$m\log\frac{d}{e} = m'\log\frac{d}{e'} \quad \text{......................(93).}$$

We shall first consider the case when $m = m'$, and therefore $e = e'$.

In this case our equations are

$$\frac{d\alpha'_n}{dt} = \frac{m}{4\pi a^2}\{\gamma'_n(L - M) - \gamma_n N\},$$

$$\frac{d\gamma'_n}{dt} = \frac{m}{4\pi a^2}\{\alpha'_n(Q - P) + \alpha_n R\},$$

$$\frac{d\alpha_n}{dt} = \frac{m}{4\pi a^2}\{\gamma_n(L - M) - \gamma'_n N\},$$

$$\frac{d\gamma_n}{dt} = \frac{m}{4\pi a^2}\{\alpha_n(Q - P) + \alpha'_n R\}.$$

Adding the first and third of these equations, we get

$$\frac{d}{dt}(\alpha'_n + \alpha_n) = \frac{m}{4\pi a^2}(L - M - N)(\gamma_n + \gamma'_n);$$

adding the second and fourth, we get

$$\frac{d}{dt}(\gamma'_n + \gamma_n) = \frac{m}{4\pi a^2}(Q + R - P)(\alpha_n + \alpha'_n).$$

Hence

$$\frac{d^2}{dt^2}(\alpha'_n + \alpha_n) + \left(\frac{m}{4\pi a^2}\right)^2 (Q + R - P)(M + N - L)(\alpha'_n + \alpha_n) = 0;$$

therefore $$\alpha'_n + \alpha_n = A\cos(\nu t + \epsilon),$$

where $$\nu^2 = \left(\frac{m}{4\pi a^2}\right)^2 (Q+R-P)(M+N-L),$$

and A and ϵ are arbitrary constants.

Substituting the values of the quantities involved in the expression for ν, we find

$$\nu^2 = \left(\frac{m}{4\pi a^2}\right)^2 4n^2(n^2-1)\left(\log\frac{8a}{d} + \log\frac{8a}{e}\right)^2;$$

therefore $$\nu = \frac{m}{2\pi a^2}\sqrt{\{n^2(n^2-1)\}}\log\frac{64a^2}{de} \qquad\ldots\ldots\ldots\ldots\ldots(94),$$

or if V be the velocity of translation of the vortex ring we have very nearly

$$\nu = \sqrt{\{n^2(n^2-1)\}}\frac{V}{a}$$

and $$\gamma'_n + \gamma_n = A\frac{\sqrt{(n^2-1)}}{n}\sin(\nu t + \epsilon)$$

Subtracting the third from the first of the four equations giving $\frac{d\alpha_n}{dt}$ &c., we get

$$\frac{d}{dt}(\alpha'_n - \alpha_n) = \frac{m}{4\pi a^2}(L+N-M)(\gamma'_n - \gamma_n).$$

Subtracting the fourth from the second of these equations, we get

$$\frac{d}{dt}(\gamma'_n - \gamma_n) = \frac{m}{4\pi a^2}(Q-P-R)(\alpha'_n - \alpha_n).$$

Hence

$$\frac{d^2}{dt^2}(\alpha'_n - \alpha_n) + \left(\frac{m}{4\pi a^2}\right)^2 (L+N-M)(R+P-Q)(\alpha'_n - \alpha_n) = 0;$$

therefore $$\alpha'_n - \alpha_n = B\cos(\mu t + \epsilon'),$$

where $$\mu^2 = \left(\frac{m}{4\pi a^2}\right)^2 (L+N-M)(R+P-Q),$$

and B and ϵ' are arbitrary constants.

Substituting the values of the quantities involved in the expression for μ^2, we find

$$\mu^2 = \left(\frac{m}{4\pi a^2}\right)^2\left(\frac{8a^2}{d^2} + (n^2 - \tfrac{3}{2})\log\frac{64a^2}{d^2} - n^2\log\frac{64a^2}{e^2}\right)$$
$$\times\left(\frac{8a^2}{d^2} + (n^2+\tfrac{1}{2})\log\frac{64a^2}{d^2} - (n^2-1)\log\frac{64a^2}{e^2}\right)$$

$$= \left(\frac{m}{2\pi a^2}\right)^2 \left\{\frac{4a^2}{d^2} - \tfrac{3}{2}\log\frac{8a}{d} - n^2 \log\frac{d}{e}\right\}$$

$$\times \left(\frac{4a^2}{d^2} + \tfrac{3}{2}\log\frac{8a}{d} - (n^2-1)\log\frac{d}{e}\right) \text{......(95)},$$

and $$\gamma'_n - \gamma_n = B' \sin(\mu t + \epsilon'),$$

where $$B' = -B\sqrt{\left[\frac{\frac{4a^2}{d^2} + \frac{3}{2}\log\frac{8a}{d} - (n^2-1)\log\frac{d}{e}}{\frac{4a^2}{d^2} - \frac{3}{2}\log\frac{8a}{d} - n^2\log\frac{d}{e}}\right]}.$$

Combining the expressions for $\alpha'_n + \alpha_n$ and $\alpha'_n - \alpha_n$, and doubling the arbitrary constants A and B for convenience, we find

$$\left.\begin{aligned}
\alpha'_n &= A\cos(\nu t + \epsilon) + B\cos(\mu t + \epsilon') \\
\alpha_n &= A\cos(\nu t + \epsilon) - B\cos(\mu t + \epsilon') \\
\gamma'_n &= \frac{\sqrt{(n^2-1)}}{n} A\sin(\nu t + \epsilon) + B'\sin(\mu t + \epsilon') \\
\gamma_n &= \frac{\sqrt{(n^2-1)}}{n} A\sin(\nu t + \epsilon) - B'\sin(\mu t + \epsilon')
\end{aligned}\right\} \text{......(96)}.$$

Since exactly the same relation exists between β'_n and δ'_n, β_n and δ_n, as between α'_n and γ'_n, α_n and γ_n, we shall have

$$\left.\begin{aligned}
\beta'_n &= C\cos(\nu t + \epsilon) + D\cos(\mu t + \epsilon') \\
\beta_n &= C\cos(\nu t + \epsilon) - D\cos(\mu t + \epsilon') \\
\delta'_n &= \frac{\sqrt{(n^2-1)}}{n} C\sin(\nu t + \epsilon) + D'\sin(\mu t + \epsilon') \\
\delta_n &= \frac{\sqrt{(n^2-1)}}{n} C\sin(\nu t + \epsilon) - D'\sin(\mu t + \epsilon')
\end{aligned}\right\} \text{......(97)},$$

where $$D' = -D\sqrt{\left[\frac{\frac{4a^2}{d^2} + \frac{3}{2}\log\frac{8a}{d} - (n^2-1)\log\frac{d}{e}}{\frac{4a^2}{d^2} - \frac{3}{2}\log\frac{8a}{d} - n^2\log\frac{d}{e}}\right]}.$$

As consequences of these equations we see (1) that the motion of the kind we have been considering is possible and stable; (2) that for each mode of displacement there are two periods of vibrations, viz. $\frac{2\pi}{\nu}$ and $\frac{2\pi}{\mu}$.

Now, if $\frac{d}{a}$ be of the same order as $\frac{e}{d}$, $\frac{a^2}{d^2}$ will be of the order $\frac{a}{e}$; and when x is large, x is very great compared with $\log x$, thus

$\frac{a}{e}$ will be great compared with $\log\frac{a}{e}$, and therefore $\frac{a^2}{d^2}$ will be great compared with $\log\frac{a}{e}$.

Thus μ will be very much greater than ν. We shall for convenience refer to the vibration expressed by $A\cos\mu t$ as the quick vibration, and to the one expressed by $A\cos\nu t$ as the slow vibration. As a very rough approximation we see that $\mu=\frac{2m}{\pi d^2}$; or the period of the vibration $=\frac{\pi^2 d^2}{m}$. This would be the period in which two infinitely long straight vortices would rotate round each other if the distance between them were great compared with the diameter of either. We also observe that the coefficients of the quick vibration in the expression for α_n and α'_n are equal in magnitude and opposite in sign, and that the same is true for the coefficients of γ_n and γ'_n. Thus, if the vortices were initially placed so that α_n was equal and opposite to α'_n, and γ_n equal and opposite to γ'_n, the slow vibrations would not be excited, and could only arise when the vortices suffered some external disturbance. This relation between α_n and α'_n, γ_n and γ'_n exists when the vortices are placed as we have supposed them, *i.e.* when they are wound round an anchor ring, the cross section of which is small compared with its aperture, and so placed that pieces of the two vortices are always at opposite extremities of a diameter of the cross section of the anchor ring.

Let us consider in more detail some of the simple cases.

(1) Let us suppose that the vortices are linked once through each other.

In this case $n=1$, and by equation (95)

$$\mu^2=\left(\frac{m}{2\pi a^2}\right)^2\left(\frac{4a^2}{d^2}-\tfrac{3}{2}\log\frac{8a}{d}-\log\frac{d}{e}\right)\left(\frac{4a^2}{d^2}+\tfrac{3}{2}\log\frac{8a}{d}\right),$$

or approximately

$$\mu=\frac{m}{\pi}\left(\frac{2}{d^2}-\frac{1}{4a^2}\log\frac{d}{e}\right).$$

(2) Let the vortices be linked twice through each other.

In this case we have approximately, since $n=2$,

$$\mu=\frac{m}{\pi}\left(\frac{2}{d^2}-\frac{7}{4a^2}\log\frac{d}{e}\right).$$

Thus this vibration is slower than the other by $\frac{3m}{4\pi^2 a^2}\log\frac{d}{e}$ vibrations in a second; this is a very small fraction of the whole number of vibrations in a second, and increases with the distance between the vortices, the cross section remaining the same.

If the vortices are linked n times through each other, we have approximately

$$\mu = \frac{m}{\pi}\left(\frac{2}{d^2} - \frac{(2n^2-1)}{4a^2}\log\frac{d}{e}\right).$$

Thus we see that the period of the vibrations gets longer as the complexity of the linking increases, but that the difference in the number of vibrations per second from this cause is small compared with the whole number of vibrations per second.

§ 43. Let us now go on to consider the case when the two vortices are of unequal strength; in this case there will for each value of a be a definite value of d, so that if the radius of the aperture of the anchor ring, on which we supposed the vortices wound, be given, the radius of the transverse section of the anchor ring will be determinate.

The relation connecting d, and e, e' the radii of the transverse sections of the vortex rings, is by equation (93)

$$m \log\frac{d}{e} = m' \log\frac{d}{e'},$$

or

$$\left(\frac{d}{e}\right)^m = \left(\frac{d}{e'}\right)^{m'}.$$

Now when a is given e, and e' are determinate, since the volume of fluid in each vortex ring remains constant; and from the equation d can be determined.

Let

$$e = re',$$
$$m = sm',$$

then we may easily prove from the above equation that

$$d = r^{\frac{1}{s-1}}(re') \qquad \text{.................. (98).}$$

Since d must be greater than re', $r^{\frac{1}{s-1}}$ must be greater than unity, thus if s be greater than unity, r must be so too, or the vortex of greatest strength must have the greatest cross section. If we are to apply our results, which were obtained on the supposition, that the distance between the vortices was great compared with the diameter of the cross section of either; we must therefore have $r^{\frac{1}{s-1}}$ large. It ought to be noticed, that r as well as s is constant, and does not depend on the radius of the rings, for if Q and Q' be the volumes of the liquid in the vortices (I) and (II) respectively, we have

$$\frac{Q}{Q'}=\frac{2\pi^2 ae^2}{2\pi^2 ae'^2}=r^2 \text{ or } r=\sqrt{\frac{Q}{Q'}} \text{ a constant quantity.}$$

With the same notation as before, we have, by equations (91) and (92),

$$\left.\begin{aligned}\frac{d\alpha'_n}{dt}&=\frac{1}{4\pi a^2}\{\gamma'_n(mL-m'M')-\gamma_n mN\}\\ \frac{d\gamma'_n}{dt}&=\frac{1}{4\pi a^2}\{\alpha'_n(m'Q'-mP)+\alpha_n mR\}\\ \frac{d\alpha_n}{dt}&=\frac{1}{4\pi a^2}\{\gamma_n(m'L-mM)-\gamma'_n m'N\}\\ \frac{d\gamma_n}{dt}&=\frac{1}{4\pi a^2}\{\alpha_n(mQ-m'P)+\alpha'_n m'R\}\end{aligned}\right\}\ldots(99).$$

If we put

$$\alpha_n=A\epsilon^{qt},\ \alpha'_n=A'\epsilon^{qt},\ \gamma_n=B\epsilon^{qt},\ \gamma'_n=B'\epsilon^{qt},$$

we find by the usual method the following equation for q:

$$\begin{aligned}q^4+\frac{1}{(4\pi a^2)^2}\{2mm'NR-(m'Q'-mP)(mL-m'M')&\\ -(mQ-m'P)(m'L-mM)\}\,q^2&\\ +\frac{1}{(4\pi a^2)^4}\{(m'Q'-mP)(mQ-m'P)-mm'R^2\}&\\ \times\{(mL-m'M')(m'L-mM)-mm'N^2\}&=0.\end{aligned}$$

If we put $q^2=-p^2$, and suppose a/d of the same order of small quantities as e/d, then we shall find, by substituting for the quantities involved, their values from equation (90), that the two values of p are approximately

$$p_1=\frac{m+m'}{\pi d^2}-\frac{mm'}{m+m'}\frac{1}{4\pi a^2}(2n^2-1)\log\frac{d^2}{ee'},$$

$$p_2=\frac{n\sqrt{(n^2-1)}}{4\pi a^2}\left\{m'\log\frac{8a}{d}+m\log\frac{8a}{e}+m\log\frac{8a}{d}+m'\log\frac{8a}{e'}\right\};$$

or, if V be the velocity of translation of the vortices, the last equation may, by the help of equation (93), be written

$$p_2=\frac{n\sqrt{(n^2-1)}\,V}{a}.$$

Thus we have, as before, a quick vibration corresponding to the root p_1, and a slow vibration corresponding to p_2.

Hence
$$\alpha_n = A\cos(p_1t+\alpha) + B\cos(p_2t+\beta),$$
$$\gamma_n = C\sin(p_1t+\alpha) + D\sin(p_2t+\beta),$$
$$\alpha'_n = A'\cos(p_1t+\alpha) + B'\cos(p_2t+\beta),$$
$$\gamma'_n = C'\sin(p_1t+\alpha) + D'\sin(p_2t+\beta),$$
where A, B, C, D, A', B', C', D' are constants, two of which are arbitrary, and the rest deducible from them, α and β are also arbitrary constants. If we substitute these values for α_n, γ_n, α'_n, γ'_n in the differential equations, we shall find that approximately
$$mA = -m'A', \qquad mC = -m'C',$$
$$A = -C, \qquad A' = -C',$$
$$B = B', \qquad D = D',$$
$$D = \frac{\sqrt{(n^2-1)}}{n}B, \qquad D' = \frac{\sqrt{(n^2-1)}}{n}B'.$$

The first four of these equations shew that, as we might have expected, the way the vortices are linked is not the same as when they are of equal strength. These equations shew that the vortex rings are now linked in the following manner.

Describe an anchor ring whose mean radius of aperture is a and the radius of whose transverse section is $\frac{m'}{m+m'}, d$, then the central line of vortex core of the vortex of strength m will always lie on the surface of this anchor ring. Describe another anchor ring with the same circular axis and the same mean radius of aperture as the first, but with a transverse section of radius $\frac{m}{m+m'} d$, then the central line of vortex core of the vortex ring, whose strength is m', will always lie on the surface of this anchor ring; and will be so situated with respect to the first vortex ring that if we take a transverse section of the anchor ring, and if C be the common centre of the two circular sections, P and Q the points where the central lines of the vortex rings cut the plane of section, then P, C, Q will be in one straight line and C will be between P and Q. If we imagine the circular axis of the anchor rings to move forward with a velocity V, and the circular axes of the vortex rings to rotate round it with angular velocity p_1, we shall get a complete representation of the motion.

§ 44. We have in the preceding investigations supposed the axes of the two vortices to be twisted r times round an anchor ring in the manner described in § 34, so that when the vortices are of equal strengths, and there are no other sinuosities in their axes, the equations to the axes of the two vortices are respectively

$$\left.\begin{aligned}\rho &= a + \tfrac{1}{2}d\cos(\mu t + r\psi)\\ z &= \mathfrak{z} + \tfrac{1}{2}d\sin(\mu t + r\psi)\end{aligned}\right\},$$

$$\left.\begin{aligned}\rho &= a - \tfrac{1}{2}d\cos(\mu t + r\psi)\\ z &= \mathfrak{z} - \tfrac{1}{2}d\sin(\mu t + r\psi)\end{aligned}\right\}.$$

When d is the diameter of the transverse section of the anchor ring, and $\mu = \frac{2m}{\pi d^2}$ approximately, its accurate value is given by equation (95), writing in that equation r in place of n.

Now, though d is small compared with a, yet it is large when compared with the quantities α_n, β_n, &c., denoting the other sinuosities; so that it is desirable to include the terms containing the squares of d in the expressions for the velocities.

By means of equations (11), (12), (14), (15) and (35), we find that the terms in d^2 in the expression for the velocity along the radius vector

$$= \frac{md^2}{8\pi a^3}\sin 2(\mu t + r\psi)\left\{2r^2\log\frac{8a}{e} - \frac{a^2}{d^2}\right\}.$$

We have only retained the largest terms in these expressions, thus we have neglected $\log\frac{8a}{d}$ in comparison with $\frac{a}{d}$.

The velocity perpendicular to the plane of the vortex by equations (17), (18) and (35), if we retain only the largest terms,

$$= \frac{md^2}{8\pi a^3}\left\{\frac{a^2}{d^2} - \tfrac{7}{32}\log\frac{64a^2}{d^2}\right\}$$
$$+ \frac{md^2}{32\pi a^3}\cos 2(\mu t + r\psi)\left(\frac{8a^2}{d^2} - (8r^2 + \tfrac{25}{4})\log\frac{64a^2}{e^2}\right).$$

We shall only require in the expression for the velocity perpendicular to the radius vector the terms containing d to the first power; we find by equations (11) and (14), that to this order of small quantities the velocity perpendicular to the radius vector

$$= -\frac{md}{4\pi a^2}\cos(\mu t + r\psi)\, r\log\frac{64a^2}{de}.$$

Now, as before, $\frac{a^2}{d^2}$ will be large compared with $\log\frac{8a}{e}$, thus we may neglect the terms involving logarithms in the expressions for the velocities.

If we go to the equation (§ 13)

$$\Sigma\left(\frac{d\alpha_r}{dt}\cos r\psi + \frac{d\beta_r}{dt}\sin r\psi\right)$$
$$- r(\alpha_r\sin r\psi - \beta_r\cos r\psi)\Psi - e\sin\chi\,.\,\mathrm{X} = \mathfrak{R},$$

and introduce into $\mathfrak{R}$ the additional terms we have just found, we get if we equate the coefficients of $\cos 2r\psi$ (neglecting the terms containing $\log \frac{64a^2}{e^2}$)

$$\frac{d\alpha_{2r}}{dt} = -\frac{m}{8\pi a}\sin 2\mu t + \frac{m}{4\pi a^2}(\gamma'_{2r}(L-M) - \gamma_{2r}N),$$

see equation (91), here L, M, N have the same values as those given by equation (90), if $2r$ be written in those equations in the place of n.

From the equation (§ 13),

$$\Sigma\left(\frac{d\gamma_r}{dt}\cos r\psi + \frac{d\delta_r}{dt}\sin r\psi\right)$$
$$- r(\gamma \sin r\psi - \delta_r \cos r\psi)\,\Psi + e\cos\chi\,.\,\mathrm{X} = w;$$

introducing into w the additional term just found, we get if we equate the terms independent of ψ on both sides of the equation,

$$\frac{d\mathfrak{z}}{dt} = \frac{m}{2\pi a}\left\{\log\frac{64a^2}{de} - 2 + \tfrac{1}{4}\right\} = \frac{m}{2\pi a}\left(\log\frac{64a^2}{de} - \tfrac{7}{4}\right);$$

if we equate the coefficient of $\cos 2r\psi$ on each side of the equation, we get, neglecting the logarithmic terms as before,

$$\frac{d\gamma_{2r}}{dt} = \frac{m}{4\pi a}\cos 2\mu t + \frac{m}{4\pi a^2}(\alpha'_{2r}(Q-P) + \alpha_{2r}R),$$

see equation (92), P, Q, R have the values given by equation (90), if $2r$ be written in those equations instead of n.

We have similarly for the other vortex

$$\frac{d\alpha'_{2r}}{dt} = -\frac{m}{8\pi a}\sin 2\mu t + \frac{m}{4\pi a^2}(\gamma_{2r}(L-M) - \gamma'_{2r}N),$$

$$\frac{d\gamma'_{2r}}{dt} = \frac{m}{4\pi a}\cos 2\mu t + \frac{m}{4\pi a^2}(\alpha_{2r}(Q-P) + \alpha'_{2r}R),$$

so $$\frac{d}{dt}(\alpha_{2r} - \alpha'_{2r}) = \frac{m}{4\pi a^2}(\gamma_{2r} - \gamma_{2r})(L+N-M),$$

and $$\frac{d}{dt}(\gamma_{2r} - \gamma'_{2r}) = \frac{m}{4\pi a^2}(\alpha_{2r} - \alpha'_{2r})(R+P-Q).$$

These equations shew that if $\alpha_{2r} = \alpha'_{2r}$ and $\gamma_{2r} = \gamma'_{2r}$ initially, they will remain equal, we shall suppose that initially these quantities are equal, so that the equation becomes

$$\frac{d\alpha_{2r}}{dt} = -\frac{m}{8\pi a}\sin 2\mu t + \frac{m}{4\pi a^2}\gamma_{2r}(L-M-N),$$

$$\frac{d\gamma_{2r}}{dt} = \frac{m}{4\pi a}\cos 2\mu t + \frac{m}{4\pi a^2}\alpha_{2r}(Q+R-P),$$

hence

$$\frac{d^2\alpha_{2r}}{dt^2} + \left(\frac{m}{4\pi a^2}\right)^2 (M+N-L)(Q+R-P)\,\alpha_{2r}$$

$$= -\frac{m}{4\pi a}\left(\mu + \frac{m}{4\pi a^2}(N+M-L)\right)\cos 2\mu t.$$

Now, by equation (90), we see that $N+M-L$ is small compared with μ, *i.e.* $\left(\frac{2m}{\pi d^2}\right)$, so that the equation is approximately

$$\frac{d^2\alpha_{2r}}{dt^2} + \nu^2\alpha_{2r} = -\frac{m\mu}{4\pi a}\cos 2\mu t,$$

if $$\nu^2 = \left(\frac{m}{4\pi a^2}\right)^2 (M+N-L)(Q+R-P).$$

Thus $\alpha_{2r} = -\frac{m\mu}{4\pi a(\nu^2 - 4\mu^2)}\cos 2\mu t +$ complementary function; equation (90) shews that ν is small compared with μ, so that we have approximately for the forced vibration

$$\alpha_{2r} = \frac{m}{16\pi a\mu}\cos 2\mu t$$

$$= \frac{d^2}{32a}\cos 2\mu t.$$

Similarly $$\gamma_{2r} = \frac{d^2}{16a}\sin 2\mu t,$$

and $$\alpha'_{2r} = \alpha_{2r},\ \gamma'_{2r} = \gamma_{2r}.$$

We thus see that any sinuosity gives rise to one of half the wave length, and that, neglecting powers of $\frac{a}{d}$ above the second, the equations to the two central lines of the vortices are

$$\left.\begin{aligned} \rho &= a\left\{1 + \frac{d}{2a}\cos(\mu t + r\psi) + \frac{d^2}{32a^2}\cos 2(\mu t + r\psi)\right\} \\ z &= \mathfrak{z} \quad + \tfrac{1}{2}d\sin(\mu t + r\psi) + \frac{d^2}{16a}\sin 2(\mu t + r\psi) \end{aligned}\right\} \ldots (100),$$

$$\left.\begin{aligned} \rho &= a\left\{1 - \tfrac{1}{2}\frac{d}{a}\cos(\mu t + r\psi) + \frac{d^2}{32a^2}\cos 2(\mu t + r\psi)\right\} \\ z &= \mathfrak{z} \quad - \tfrac{1}{2}d\sin(\mu t + r\psi) + \frac{d^2}{16a}\sin 2(\mu t + r\psi) \end{aligned}\right\} \ldots (101).$$

We have thus proved that two vortices linked in the manner described in § (34), are capable of steady motion, and that this steady motion is stable; and if the vortices are of equal strength, their central lines take the shapes given by equations (100) and (101).

§ 45. We may prove that if the vortices are to retain an approximately circular form, they must be linked in this manner; for Sir W. Thomson has pointed out that the condition for steady motion is that, with a given force resultant of the impulse and given vorticity, the kinetic energy must be a maximum or a minimum. If we imagine two approximately circular equal vortices linked through each other, so that the distance between their centres is considerable compared with the radius of the rings; then if we give one of them a motion of translation so as to make its centre approach that of the other, we increase the kinetic energy without altering the impulse or vorticity; thus, when the centres are not near together, the kinetic energy is not a maximum or minimum, and thus the motion cannot be steady, and when the centres are close together, the motion is evidently as we have described it.

§ 46. The force resultant of the impulse, and the resultant moment of momentum, remain constant as long as the motion of the linked vortices is not disturbed by external circumstances (see §§ 4, 5), they will thus be constants determining the size of the system. We can express a and d in terms of I the force resultant of the impulse, and Γ the resultant moment of momentum. First, suppose the vortices are of equal strength, then by § 5

$$I = 4m\pi a^2\rho \text{ .... } (102),$$

ρ being the density of the fluid.

This equation gives a the radius of the aperture of the vortices.

By equation (3),

$$\Gamma = m\rho \int (x^2 + y^2) \frac{dz}{d\psi} d\psi,$$

from the equation to the circular axis of the ring

$$\frac{dz}{d\psi} = \tfrac{1}{2} rd \cos(r\psi + \mu t),$$

and $$x^2 + y^2 = a^2 + ad\cos(r\psi + \mu t) + \frac{d^2}{4}\cos^2(r\psi + \mu t);$$

integrating all round the ring, we find that the moment of momentum for each ring $= \tfrac{1}{2} m\rho\pi rad^2$, so that

$$\Gamma = \pi m\rho rad^2 \text{ .... } (103).$$

This is the same expression for the resultant moment of momentum as that given by Sir William Thomson in his paper on "Vortex Statics," *Phil. Mag.*, Aug. 1881, p. 102.

From equations (102) and (103) we find

$$\frac{d^2}{a^2} = 4\frac{\Gamma\,(4\pi m\rho)^{\frac{1}{2}}}{rI^{\frac{3}{2}}}.$$

Now $\dfrac{d^2}{a^2}$ is small, hence the condition that the rings should be approximately circular and the motion steady, is that $(4\pi m\rho)^{\frac{1}{2}}\dfrac{\Gamma}{I^{\frac{3}{2}}}$ should be small.

If the vortices instead of being equal are of strengths m and m', we find in a similar way

$$\left.\begin{aligned} I &= 2\pi a^2\rho\,(m+m'),\\ \Gamma &= 2\pi\rho a d^2\,\frac{rmm'}{m+m'}\end{aligned}\right\}\text{.....................}(104),$$

these equations determine a and d, but equation (98) determines d when a is known; hence, unless the value of d, determined from equations (104), differs but infinitesimally from that determined by equation (98), the motion cannot be steady if the vortex rings are nearly circular.

From equation (98) we see that

$$2\pi^2 a d^2 = Q^{\frac{m}{m-m'}}\,Q'^{\frac{m'}{m'-m}},$$

when Q and Q' are the volumes of vortices (I) and (II) respectively; hence, from equation (104), we get

$$\Gamma = \frac{\rho r}{\pi}\,Q^{\frac{m}{m-m'}}\,Q'^{\frac{m'}{m'-m}}\,\frac{mm'}{m+m'}\text{.....................}(105).$$

Thus when two unequal vortices are linked together, unless the moment of momentum has the value given by this equation, the vortices cannot be linked in the manner described in § (38), and so there can be no steady motion with the vortex rings approximately circular.

§ 47. In the case we have been considering only two vortices were linked together; we can, however, apply the same method to the case when any number of equal vortices are twisted round each other. We must suppose the vortices linked round an anchor ring whose transverse section is small compared with its aperture, the vortices being arranged so that the circular lines of their vortex cores cut the plane of any transverse section in the angular points of a regular polygon inscribed in the circular transverse section of the anchor ring. The question however arises whether such a distribution of vortices would be stable, and the following considerations will shew, I think, that there must be a limit to the number of vortices which can be in stable equilibrium when arranged at equal intervals round the circumference of a circle. When we have a large number of vortices arranged round the circle the distance between consecutive ones must be small compared with the radius of the circle, and thus, as the number of vortices increases, the system will approximate to a cylindrical vortex sheet; but for this case there is discontinuity in the motion when we pass from the inside to the outside of the cylinder, and, as Sir William Thomson has proved, all discontinuous fluid motion is unstable; the equilibrium in this case would be unstable and the vortices would probably break up into separate groups, each group consisting of a comparatively small number of vortices.

We shall now go on to investigate the number of vortices which can be arranged in the manner described and yet be in stable equilibrium.

As the distance between the vortices is small compared with the radius of their apertures this problem will be very approximately the same as if the same number of infinitely long straight vortices were arranged at equal intervals round the circumference of a circle, and as the mathematical work in this case is much simpler than in the original one, it is the case we shall consider in the subsequent investigation.

§ 48. The problem we are about to investigate is this: A system consisting of n equal straight cylindrical vortices, arranged at equal intervals round the circumference of a circle, is slightly displaced; what is the subsequent motion? We suppose the radius of a cross section of a vortex to be small compared with the distance between two vortices.

Take as the origin of coordinates the centre of gravity of the vortices in their undisturbed position. Let the position of the sth vortex be determined by the radius vector $(r + x_s)$, say r_s, and the angle $(\omega_s + \theta_s)$, say ϕ_s when r and ω_s are the coordinates of the position the sth vortex would occupy if the motion were undisturbed. Thus, x_s and θ_s will be small quantities.

Let the strength of each of the vortices be m.

The stream function due to a single vortex of strength m at a point whose distance from the vortex is ρ

$$= -\frac{m}{\pi} \log \rho.$$

Thus ψ, the stream function at a point R and ϕ due to the n vortices, is given by the equation

$$\psi = -\frac{m}{2\pi}\Sigma \log \{r^2_s + R^2 - 2r_s R \cos(\phi_s - \phi)\}.$$

The velocity along the radius vector

$$= \frac{d\psi}{Rd\phi}$$

$$= -\frac{m}{\pi}\Sigma \frac{r_s \sin(\phi - \phi_s)}{\{r_s^2 + R^2 - 2r_s R\cos(\phi - \phi_s)\}}.$$

The velocity perpendicular to the radius vector $= -\frac{d\psi}{dR}$

$$= \frac{m}{\pi}\Sigma \frac{\{R - r_s \cos(\phi - \phi_s)\}}{\{r_s^2 + R^2 - 2r_s R\cos(\phi - \phi_s)\}}.$$

Now let the point (R, ϕ) coincide in position with the sth vortex, then the velocity along the radius vector

$$= -\frac{m}{\pi}\left\{\frac{r_1 \sin(\phi_s - \phi_1)}{r_1^2 + r_s^2 - 2r_1 r_s \cos(\phi_s - \phi_1)} + \frac{r_2 \sin(\phi_s - \phi_2)}{r_2^2 + r_s^2 - 2r_2 r_s \cos(\phi_s - \phi_2)} + \ldots\right\} \quad \ldots\ldots\ldots\ldots(105^*).$$

The velocity perpendicular to the radius vector

$$= \frac{m}{\pi}\left\{\frac{r_s - r_1 \cos(\phi_s - \phi_1)}{r_1^2 + r_s^2 - 2r_1 r_s \cos(\phi_s - \phi_1)} + \frac{r_s - r_2 \cos(\phi_s - \phi_2)}{r_2^2 + r_s^2 - 2r_2 r_s \cos(\phi_s - \phi_2)} + \ldots\right\}.$$

In the undisturbed position the vortex has no velocity along the radius vector, so that its radial velocity will be a function of the quantities x_m and θ_m, which vanishes when they do, we proceed to find this function. Calling the radial velocity when expressed as in equation (105*) v, the radial velocity at the sth vortex

$$= \frac{dv}{dr_1}x_1 + \frac{dv}{dr_2}x_2 + \ldots + \frac{dv}{d\phi_1}\theta_1 + \frac{dv}{d\phi_2}\theta_2 + \ldots,$$

neglecting the squares of the small quantities x and θ, and after differentiation putting $r_1 = r_2 = r$ and $\theta_1 = \theta_2 = \ldots = 0$.

Now
$$\frac{d}{dr_s}\frac{r_1 \sin(\phi_s - \phi_1)}{r_s^2 + r_1^2 - 2r_1 r_s \cos(\phi_s - \phi_1)}$$

$$= -\frac{2r_1 \sin(\phi_s - \phi_1)\{r_s - r_1\cos(\phi_s - \phi_1)\}}{\{r_s^2 + r_1^2 - 2r_1 r_s \cos(\phi_s - \phi_1)\}^2},$$

and when $r_1 = r_2 = r$ and $\theta = 0$, this becomes

$$-\frac{\sin(\omega_s - \omega_1)}{2r^2\{1 - \cos(\omega_s - \omega_1)\}},$$

Again

$$\frac{d}{dr_1}\frac{r_1 \sin(\phi_s-\phi_1)}{r_s^2+r_1^2-2r_1r_s\cos(\phi_s-\phi_1)}=\frac{\sin(\phi_s-\phi_1)}{r_s^2+r_1^2-2r_sr_1\cos(\phi_s-\phi_1)}$$
$$-\frac{2r_1\sin(\phi_s-\phi_1)\{r_1-r_s\cos(\phi_s-\phi_1)\}}{\{r_s^2+r_1^2-2r_1r_s\cos(\phi_s-\phi_1)\}^2},$$

and when all the r's are put equal to r, and all the θ's zero, this becomes

$$\frac{\sin(\omega_s-\omega_1)}{2r^2\{1-\cos(\omega_s-\omega_1)\}}-\frac{\sin(\omega_s-\omega_1)}{2r^2\{1-\cos(\omega_s-\omega_1)\}}=0.$$

If ψ be written for a moment for $\phi_s-\phi_1$,

$$\frac{d}{d\psi}\frac{r_1\sin\psi}{r_1^2+r_s^2-2r_1r_s\cos\psi}=\frac{r_1\cos\psi}{(r_1^2+r_s^2-2r_1r_s\cos\psi)}$$
$$-\frac{2r_1^2r_s\sin^2\psi}{(r_1^2+r_s^2-2r_1r_s\cos\psi)^2};$$

when $r_1=r_s=r$ and $\psi=\omega_s-\omega_1$, this becomes

$$-\frac{1}{2r\{1-\cos(\omega_s-\omega_1)\}},$$

hence the velocity along the radius vector of the sth vortex

$$=\frac{m}{2\pi r}\left\{\frac{\theta_s-\theta_1}{1-\cos(\omega_s-\omega_1)}+\frac{\theta_s-\theta_2}{1-\cos(\omega_s-\omega_2)}+\dots\right\}\dots(106).$$

The coefficient of x_s vanishes, because it is evidently proportional to the expression for the velocity along the radius vector in the undisturbed motion, which we know vanishes.

The radial velocity of the sth vortex also $=\frac{dx_s}{dt}$; therefore

$$\frac{dx_s}{dt}=\frac{m}{2\pi r}\left\{\frac{\theta_s-\theta_1}{1-\cos(\omega_s-\omega_1)}+\frac{\theta_s-\theta_2}{1-\cos(\omega_s-\omega_2)}+\dots\right\}\dots(107).$$

We proceed to find the expression for the velocity perpendicular to the radius vector.

Now

$$\frac{d}{dr_s}\frac{r_s-r_1\cos(\phi_s-\phi_1)}{r_s^2+r_1^2-2r_1r_s\cos(\phi_s-\phi_1)}=\frac{1}{r_s^2+r_1^2-2r_1r_s\cos(\phi_s-\phi_1)}$$
$$-\frac{2\{r_s-r_1\cos(\phi_s-\phi_1)\}^2}{\{r_1^2+r_s^2-2r_1r_s\cos(\phi_s-\phi_1)\}^2};$$

when $r_1=r=r_s$ and $\phi_s=\omega_s$, $\phi_1=\omega_1$, this becomes

$$\frac{1}{2r^2}\frac{\cos(\omega_s-\omega_1)}{1-\cos(\omega_s-\omega_1)},$$

$$\frac{d}{dr_1}\frac{\{r_s - r_1\cos(\phi_s - \phi_1)\}}{r_1^2 + r_s^2 - 2r_1r_s\cos(\phi_s - \phi_1)} = -\frac{\cos(\phi_s - \phi_1)}{r_1^2 + r_s^2 - 2r_1r_s\cos(\phi_s - \phi_1)}$$
$$-\frac{2\{r_s - r_1\cos(\phi_s - \phi_1)\}\{r_1 - r_s\cos(\phi_s - \phi_1)\}}{\{r_1^2 + r_s^2 - 2r_1r_s\cos(\phi_s - \phi_1)\}^2},$$

when $r_1 = r_s = r$, &c., this becomes

$$-\frac{1}{2r^2\{1 - \cos(\omega_s - \omega_1)\}}.$$

Writing as before ψ for $\phi_s - \phi_1$,

$$\frac{d}{d\psi}\frac{r_s - r_1\cos\psi}{r_1^2 + r_s^2 - 2r_1r_s\cos\psi} = \frac{r_1\sin\psi}{r_1^2 + r_s^2 - 2r_1r_s\cos\psi}$$
$$-\frac{2r_1r_s\sin\psi\,(r_s - r_1\cos\psi)}{(r_1^2 + r_1^2 - 2r_1r_s\cos\psi)^2},$$

and when $r_1 = r_s = r$, &c. this vanishes.

Thus the increment in the velocity perpendicular to the radius vector

$$= -\frac{m}{2\pi r^2}\left\{-x_s\left(\frac{\cos(\omega_s - \omega_1)}{1 - \cos(\omega_s - \omega_1)} + \frac{\cos(\omega_s - \omega_2)}{1 - \cos(\omega_s - \omega_2)} + \dots\right)\right.$$
$$\left. + \frac{x_1}{1 - \cos(\omega_s - \omega_1)} + \frac{x_2}{1 - \cos(\omega_s - \omega_2)} + \dots\right\} \quad \text{............ (108).}$$

But if Ω be the angular velocity of revolution of the system when the motion is steady, the increment in the velocity perpendicular to the radius vector $= r\dfrac{d\theta_s}{dt} + x_s\Omega$. Now

$$\Omega = \frac{m}{2\pi r^2}(n - 1),$$

hence

$$\frac{rd\theta_s}{dt} = -\frac{m}{2\pi r^2}\left\{-x_s\left[\frac{\cos(\omega_s - \omega_1)}{1 - \cos(\omega_s - \omega_1)} + \frac{\cos(\omega_s - \omega_2)}{1 - \cos(\omega_s - \omega_2)} + \dots - (n - 1)\right]\right.$$
$$\left. + \frac{x_1}{1 - \cos(\omega_s - \omega_1)} + \frac{x_2}{1 - \cos(\omega_s - \omega_2)} + \dots\right\} \quad \text{............ (109).}$$

Now we can prove by Trigonometry that

$$\frac{\cos(\omega_s - \omega_1)}{1 - \cos(\omega_s - \omega_1)} + \frac{\cos(\omega_s - \omega_2)}{1 - \cos(\omega_s - \omega_2)} + \dots = \frac{(n - 1)(n - 5)}{6} \quad \text{...(110).}$$

Substituting this value, we find

$$\frac{rd\theta_s}{dt} = -\frac{m}{2\pi r^2}\left(-\frac{x_s(n - 1)(n - 11)}{6} + \frac{x_1}{1 - \cos(\omega_s - \omega_1)}\right.$$
$$\left. + \frac{x_2}{1 - \cos(\omega_s - \omega_2)} + \dots\right) \quad \text{...(111).}$$

We have also

$$\theta_1 + \theta_2 + \theta_3 + \ldots \theta_n = 0;$$

and since $\Sigma m r^2$ is constant,

$$x_1 + x_2 + x_3 + \ldots x_n = 0.$$

These equations and equation (107) will enable us to solve the problem we are considering.

Let us apply them to the various cases in succession.

Three Vortices.

§ 49. The three vortices were originally placed at the angular points of an equilateral triangle, hence

$$\omega_2 - \omega_1 = 120°,$$
$$\omega_3 - \omega_2 = 120°;$$

therefore by equation (107),

$$\frac{dx_1}{dt} = \frac{m}{2\pi r}\left(\frac{\theta_1 - \theta_2}{\frac{3}{2}} + \frac{\theta_1 - \theta_3}{\frac{3}{2}}\right),$$

and

$$\theta_1 + \theta_2 + \theta_3 = 0;$$

therefore

$$\left.\begin{aligned} \frac{dx_1}{dt} &= \frac{m}{\pi r}\theta_1 \\ \frac{dx_2}{dt} &= \frac{m}{\pi r}\theta_2 \\ \frac{dx_3}{dt} &= \frac{m}{\pi r}\theta_3 \end{aligned}\right\} \ldots\ldots\ldots\ldots\ldots\ldots\ldots\ldots (112).$$

similarly

By equation (111), we have

$$\frac{r d\theta_1}{dt} = -\frac{m}{2\pi r^2}\left\{\frac{8}{3}x_1 + \frac{x_2}{\frac{3}{2}} + \frac{x_3}{\frac{3}{2}}\right\};$$

or since

$$x_1 + x_2 + x_3 = 0,$$

$$\left.\begin{aligned} \frac{r d\theta_1}{dt} &= -\frac{m}{\pi r^2}x_1 \\ \frac{r d\theta_2}{dt} &= -\frac{m}{\pi r^2}x_2 \\ \frac{r d\theta_3}{dt} &= -\frac{m}{\pi r^2}x_3 \end{aligned}\right\} \ldots\ldots\ldots\ldots\ldots\ldots\ldots\ldots (113),$$

similarly

hence

$$\frac{d^2 x_1}{dt^2} + \frac{m^2}{\pi^2 r^4}x_1 = 0;$$

therefore

$$x_1 = A \sin\left(\frac{m}{\pi r^2} t + a\right) \text{ and } r\theta_1 = A \cos\left(\frac{m}{\pi r^2} t + a\right),$$

with similar expressions for x_2 and x_3, $r\theta_2$ and $r\theta_3$.

Thus the motion in this case is stable, and the time of a small oscillation $= \dfrac{2\pi^2 r^2}{m}$, the same as the period of rotation of the undisturbed system.

Four Vortices.

§ 50. Let us suppose that the four vortices are initially at the angular points of a square.

Our equations in this case are

$$\frac{dx_1}{dt} = \frac{m}{2\pi r}\left(\frac{\theta_1 - \theta_2}{1} + \frac{\theta_1 - \theta_3}{2} + \frac{\theta_1 - \theta_4}{1}\right);$$

or since $\theta_1 + \theta_2 + \theta_3 + \theta_4 = 0$,

$$\left.\begin{aligned} \frac{dx_1}{dt} &= \frac{m}{4\pi r}(7\theta_1 + \theta_3) \\ \frac{dx_2}{dt} &= \frac{m}{4\pi r}(7\theta_2 + \theta_4) \\ \frac{dx_3}{dt} &= \frac{m}{4\pi r}(7\theta_3 + \theta_1) \\ \frac{dx_4}{dt} &= \frac{m}{4\pi r}(7\theta_4 + \theta_2) \end{aligned}\right\} \quad \text{..................(114).}$$

Equation (111) gives

$$\frac{rd\theta_1}{dt} = -\frac{m}{2\pi r^2}\left(\tfrac{3}{2} x_1 + x_2 + \tfrac{1}{2} x_3 + x_4\right);$$

therefore, similarly

$$\left.\begin{aligned} \frac{rd\theta_1}{dt} &= -\frac{m}{4\pi r^2}(5x_1 - x_3) \\ \frac{rd\theta_2}{dt} &= -\frac{m}{4\pi r^2}(5x_2 - x_4) \\ \frac{rd\theta_3}{dt} &= -\frac{m}{4\pi r^2}(5x_3 - x_1) \\ \frac{rd\theta_4}{dt} &= -\frac{m}{4\pi r^2}(5x_4 - x_2) \end{aligned}\right\} \quad \text{................(115).}$$

Thus

$$\frac{d}{dt}(x_1 + x_3) = \frac{2m}{\pi r}(\theta_1 + \theta_3),$$

and $$\frac{d}{dt}(\theta_1 + \theta_3) = -\frac{m}{\pi r^3}(x_1 + x_3);$$

hence $$\frac{d^2}{dt^2}(x_1 + x_3) + \frac{2m^2}{\pi^2 r^4}(x_1 + x_3) = 0;$$

therefore $$x_1 + x_3 = 2A \cos\left(\frac{m}{\pi r^2}\sqrt{2}t + \beta\right).$$

Similarly we may prove

$$\frac{d^2}{dt^2}(x_1 - x_3) + \frac{9m^2}{4\pi^2 r^4}(x_1 - x_3) = 0;$$

therefore $$x_1 - x_3 = 2A' \cos\left(\frac{3m}{2\pi r^2}t + \gamma\right);$$

therefore

$$\left.\begin{aligned} x_1 &= A\cos\left(\frac{m}{\pi r^2}\sqrt{2}t + \beta\right) + A'\cos\left(\frac{3m}{2\pi r^2}t + \gamma\right) \\ x_3 &= A\cos\left(\frac{m}{\pi r^2}\sqrt{2}t + \beta\right) - A'\cos\left(\frac{3m}{2\pi r^2}t + \gamma\right) \end{aligned}\right\} \quad \ldots\ldots(116),$$

with corresponding expressions for x_2 and x_4, and

$$r\theta_1 = -\frac{A}{\sqrt{2}}\sin\left(\frac{m}{\pi r^2}\sqrt{2}t + \beta\right) - A'\sin\left(\frac{3m}{2\pi r^2}t + \gamma\right) \ldots\ldots(117),$$

with similar expressions for θ_2, θ_3, θ_4.

These equations shew that the motion is stable, and that the periods of the vibration are $\frac{2\pi^2 r^2}{m\sqrt{2}}$ and $\frac{4\pi^2 r^2}{3m}$. Both of these periods are smaller than the period for three vortices.

Five Vortices.

§ 51. Let us suppose that the five vortices are initially at the angular points of a regular pentagon.

Then by equation (107), we have

$$\frac{dx_1}{dt} = \frac{m}{2\pi r}\left(\frac{\theta_1 - \theta_2}{1 - \cos\frac{2}{5}\pi} + \frac{\theta_1 - \theta_3}{1 - \cos\frac{4}{5}\pi} + \frac{\theta_1 - \theta_4}{1 - \cos\frac{6}{5}\pi} + \frac{\theta_1 - \theta_5}{1 - \cos\frac{8}{5}\pi}\right).$$

Now we can prove by Trigonometry that

$$\frac{1}{1 - \cos\frac{2\pi}{n}} + \frac{1}{1 - \cos\frac{4\pi}{n}} + \ldots \frac{1}{1 - \cos\frac{2(n-1)\pi}{n}} = \frac{n^2 - 1}{6};$$

hence

$$\frac{1}{1-\cos\frac{2}{5}\pi}+\frac{1}{1-\cos\frac{4}{5}\pi}+\frac{1}{1-\cos\frac{6}{5}\pi}+\frac{1}{1-\cos\frac{8}{5}\pi}=4,$$

and

$$\frac{dx_1}{dt}=\frac{m}{2\pi r}\left\{\theta_1\left(4-\frac{1}{1-\cos\frac{4}{5}\pi}\right)+(\theta_2+\theta_5)\left(\frac{1}{1-\cos\frac{4}{5}\pi}-\frac{1}{1-\cos\frac{2}{5}\pi}\right)\right\};$$

or if
$$a=4+\frac{1}{1-\cos\frac{4}{5}\pi}=\frac{25-\sqrt{5}}{5},$$

$$b=\frac{1}{1-\cos\frac{4}{5}\pi}-\frac{1}{1-\cos\frac{2}{5}\pi}=-\frac{2}{\sqrt{5}},$$

$$\frac{dx_1}{dt}=\frac{m}{2\pi r}\{a\theta_1+b(\theta_2+\theta_5)\},$$

with symmetrical expressions for $\frac{dx_2}{dt}$, $\frac{dx_3}{dt}$, &c.

By equation (111),

$$\frac{d\theta_1}{dt}=-\frac{m}{2\pi r^3}\left(4x_1+\frac{x_2}{1-\cos\frac{2}{5}\pi}+\frac{x_3}{1-\cos\frac{4}{5}\pi}+\frac{x_4}{1-\cos\frac{6}{5}\pi}+\frac{x_5}{1-\cos\frac{8}{5}\pi}\right)$$

$$=-\frac{m}{2\pi r^3}\left\{x_1\left(4-\frac{1}{1-\cos\frac{4}{5}\pi}\right)+(x_2+x_5)\left(\frac{1}{1-\cos\frac{2}{5}\pi}-\frac{1}{1-\cos\frac{4}{5}\pi}\right)\right\};$$

or if
$$c=4-\frac{1}{1-\cos\frac{4}{5}\pi}=\frac{15+\sqrt{5}}{5},$$

$$\frac{d\theta_1}{dt}=-\frac{m^2}{2\pi r^3}\{cx_1-b(x_2+x_5)\}.$$

Hence

$$\frac{d^2x_1}{dt^2}=-\frac{m^2}{4\pi^2r^4}\{x_1(ac-b^2)+(x_2+x_5)(bc-ab+b^2)\}\ldots\ldots(118),$$

say
$$\frac{d^2x_1}{dt^2}=-\{a'x_1+b'(x_2+x_5)\},$$

with symmetrical expressions for x_2, x_3, &c.

If x_1, x_2... vary as $e^{\lambda t}$, the equation to determine λ is

$$\begin{vmatrix} a'+\lambda^2, & b', & 0, & 0, & b' \\ b', & a'+\lambda^2, & b', & 0, & 0 \\ 0, & b', & a'+\lambda^2, & b', & 0 \\ 0, & 0, & b', & a'+\lambda^2, & b' \\ b', & 0, & 0, & b', & a'+\lambda^2 \end{vmatrix}=0.$$

Now this determinant is of the form

$$\begin{vmatrix} a_1, & a_2, & \ldots a_n \\ a_n, & a_1, & \ldots a_{n-1} \\ a_{n-1}, & a_n, & \ldots a_{n-2} \\ \ldots & \ldots & \ldots \\ a_2, & a_3, & \ldots a_1 \end{vmatrix},$$

which equals

$$(a_1 + a_2 + \ldots + a_n)\, \Pi\, (a_1 + a_2\omega + a_3\omega^2 + a_n\omega^{n-1}),$$

where ω is one of the roots of the equation $x^n - 1 = 0$, unity being excepted (Scott's *Treatise on Determinants*, p. 82).

Thus, if $1, \omega, \omega^2, \omega^3, \omega^4$ are the fifth roots of unity, the determinant we are concerned with splits up into

$$(a' + \lambda^2 + 2b')\,(a' + \lambda^2 + \omega b' + \omega^4 b')\,(a' + \lambda^2 + \omega^2 b' + \omega^8 b')$$
$$\times\,(a' + \lambda^2 + \omega^3 b' + \omega^{12} b')\,(a' + \lambda^2 + \omega^4 b' + \omega^{16} b')$$
$$= (a' + \lambda^2 + 2b')\,\{a' + \lambda^2 + b'(\omega + \omega^4)\}^2\,\{a' + \lambda^2 + b'(\omega^2 + \omega^3)\}^2.$$

Now

$$\omega = \cos \tfrac{2}{5}\pi + i \sin \tfrac{2}{5}\pi = \cos 72^\circ + i \sin 72^\circ,$$
$$\omega^2 = \cos \tfrac{4}{5}\pi + i \sin \tfrac{4}{5}\pi = \cos 144^\circ + i \sin 144^\circ,$$
$$\omega^3 = \cos \tfrac{6}{5}\pi + i \sin \tfrac{6}{5}\pi = \cos 144^\circ - i \sin 144^\circ,$$
$$\omega^4 = \cos \tfrac{8}{5}\pi + i \sin \tfrac{8}{5}\pi = \cos 72^\circ - i \sin 72^\circ.$$

Thus the equation to determine λ^2 becomes

$$(a' + \lambda^2 + 2b')(a' + \lambda^2 + 2 \cos 72^\circ b')^2 (a' + \lambda^2 + 2 \cos 144^\circ b')^2 = 0.$$

It can be proved in exactly the same way as in the corresponding case of a material system (Thomson and Tait's *Natural Philosophy*, § 343. m), that equal roots will not introduce terms of the form $te^{\lambda t}$ into the solution. Thus the sole condition of stability is, that the values of λ^2 should all be negative.

The values of λ^2 are

$$-(a' + 2b')$$
$$-(a' + 2 \cos 72^\circ b')$$
$$-(a' + 2 \cos 144^\circ b'),$$

where
$$a' = \frac{m^2}{4\pi^2 r^4}(ac - b^2) \qquad = \frac{m^2}{4\pi^2 r^4}\,\frac{2\,(35 + \sqrt{5})}{5},$$
$$b' = \frac{m^2}{4\pi^2 r^4}(bc - ab + b^2) = \frac{m^2}{\pi^2 r}\,\frac{1}{\sqrt{5}}$$

therefore the values of λ^2 are

$$-\frac{m^2}{2\pi^2 r^4}\left(\frac{35+\sqrt{5}}{5}+\frac{4\sqrt{5}}{5}\right)=-\frac{m^2}{2\pi^2 r^4}(7+\sqrt{5})$$

$$-\frac{m^2}{2\pi^2 r^4}\left(\frac{35+\sqrt{5}}{5}+\frac{\sqrt{5}-1}{\sqrt{5}}\right)=-\frac{m^2}{2\pi^2 r^4}8$$

$$-\frac{m^2}{2\pi^2 r^4}\left(\frac{35+\sqrt{5}}{5}-\frac{(1+\sqrt{5})}{\sqrt{5}}\right)=-\frac{m^2}{2\pi^2 r^4}6.$$

Thus all the values of λ^2 are negative, and the periods of vibration are

$$\frac{4\pi^2 r^2}{m\sqrt{(14+2\sqrt{5})}},$$

$$\frac{\pi^2 r^2}{m},$$

$$\frac{2\pi^2 r^2}{m\sqrt{3}}.$$

Six Vortices.

§ 52. Let us suppose that the vortices are arranged at the angular points of a regular hexagon, then, using the same notation as before, we have by equation (107)

$$\frac{dx_1}{dt}=\frac{m}{2\pi r}\left\{\frac{\theta_1-\theta_2}{1-\cos 60^\circ}+\frac{\theta_1-\theta_3}{1-\cos 120^\circ}+\frac{\theta_1-\theta_4}{1-\cos 180^\circ}\right.$$
$$\left.+\frac{\theta_1-\theta_5}{1-\cos 240^\circ}+\frac{\theta_1-\theta_6}{1-\cos 300^\circ}\right\},$$

or since $\theta_1+\theta_2+\theta_3+\theta_4+\theta_5+\theta_6=0$,

$$\frac{dx_1}{dt}=\frac{m}{2\pi r}\{\tfrac{13}{2}\theta_1-\tfrac{4}{3}(\theta_2+\theta_6)+\tfrac{1}{6}\theta_4\}\text{ ......(118).}$$

Again, by equation (111),

$$\frac{rd\theta_1}{dt}=-\frac{m}{2\pi r^3}\left\{\tfrac{25}{6}x_1+\frac{x_2}{1-\cos 60^\circ}+\frac{x_3}{1-\cos 120^\circ}+\frac{x_4}{1-\cos 180^\circ}\right.$$
$$\left.+\frac{x_5}{1-\cos 240^\circ}+\frac{x_6}{1-\cos 300^\circ}\right\},$$

since $x_1+x_2+x_3+x_4+x_5+x_6=0$,

$$\frac{rd\theta_1}{dt}=-\frac{m}{2\pi r^3}\{\tfrac{7}{2}x_1+\tfrac{4}{3}(x_2+x_6)-\tfrac{1}{6}x_4\}\text{ ...... (119).}$$

By means of equations (118) and (119), we get

$$\frac{d^2x_1}{dt^2} = -\frac{m^2}{(2\pi r^2)^2}\left\{\tfrac{41}{2}x_1 + \tfrac{16}{3}(x_2 + x_6) + \tfrac{5}{8}x_4\right\}.$$

Say $$\frac{d^2x_1}{dt^2} = -\{\alpha x_1 + \beta(x_2 + x_6) + \gamma x_4\} \quad \text{.........} \ (120),$$

with similar equations for $x_2 \dots x_6$.

Thus, if $x_1 \dots x_6$ vary as $\epsilon^{\lambda t}$, the equation to determine λ is

$$\begin{vmatrix} \lambda^2+\alpha, & \beta, & 0, & \gamma, & 0, & \beta \\ \beta, & \lambda^2+\alpha, & \beta, & 0, & \gamma, & 0 \\ 0, & \beta, & \lambda^2+\alpha, & \beta, & 0, & \gamma \\ \gamma, & 0, & \beta, & \lambda^2+\alpha, & \beta, & 0 \\ 0, & \gamma, & 0, & \beta, & \lambda^2+\alpha, & \beta \\ \beta, & 0, & \gamma, & 0, & \beta, & \lambda^2+\alpha \end{vmatrix} = 0.$$

If 1, ω, ω^2, ω^3, ω^4, ω^5 are the sixth roots of unity, this equation splits up into the factors

$$\begin{aligned} &\lambda^2 + \alpha + 2\beta + \gamma, \\ &\lambda^2 + \alpha + \omega\beta + \omega^3\gamma + \omega^5\beta, \\ &\lambda^2 + \alpha + \omega^2\beta + \omega^6\gamma + \omega^4\beta, \\ &\lambda^2 + \alpha + \omega^3\beta + \omega^3\gamma + \omega^3\beta, \\ &\lambda^2 + \alpha + \omega^4\beta + \gamma + \omega^2\beta, \\ &\lambda^2 + \alpha + \omega^5\beta + \omega^3\gamma + \omega\beta. \end{aligned}$$

The second and last of these factors are the same, and so are also the third and fifth; thus the values of λ^2 are

$$\left.\begin{aligned} \lambda^2 &= -(\alpha + 2\beta + \gamma) = -\frac{m^2}{(2\pi r^2)^2}\,32 \\ \lambda^2 &= -(\alpha + \beta - \gamma) = -\frac{m^2}{(2\pi r^2)^2}\,25 \\ \lambda^2 &= -(\alpha - \beta + \gamma) = -\frac{m^2}{(2\pi r^2)^2}\,16 \\ \lambda^2 &= -(\alpha - 2\beta - \gamma) = -\frac{m^2}{(2\pi r^2)^2}\,9 \end{aligned}\right\} \text{............} (121).$$

Thus the values of λ^2 are all negative, and, as before, equal roots do not affect the stability, so that the steady motion is stable, the times of oscillations are

$$\frac{4\pi^2 r^2}{m\sqrt{(32)}}, \quad \frac{4\pi^2 r^2}{m\,5}, \quad \frac{4\pi^2 r^2}{m\,4}, \quad \frac{4\pi^2 r^2}{m\,3}.$$

Seven Vortices.

§ 53. Let us suppose that the vortices are arranged at equal intervals round the circumference of a circle, then using the same notation as before, we have, by equation (107),

$$\frac{dx_1}{dt}=\frac{m}{2\pi r}\left\{\frac{\theta_1-\theta_2}{1-\cos\frac{2}{7}\pi}+\frac{\theta_1-\theta_3}{1-\cos\frac{4}{7}\pi}+\frac{\theta_1-\theta_4}{1-\cos\frac{6}{7}\pi}\right.$$

$$\left.+\frac{\theta_1-\theta_5}{1-\cos\frac{8}{7}\pi}+\frac{\theta_1-\theta_6}{1-\cos\frac{10}{7}\pi}+\frac{\theta_1-\theta_7}{1-\cos\frac{12}{7}\pi}\right\},$$

or since $\theta_1+\theta_2+\theta_3+\dots\theta_7=0,$

and

$$\frac{1}{1-\cos\frac{2}{7}\pi}+\frac{1}{1-\cos\frac{4}{7}\pi}+\dots\frac{1}{1-\cos\frac{12}{7}\pi}=8 \text{ (see equation 110)},$$

$$\frac{dx_1}{dt}=\frac{m}{2\pi r}\left\{\left(8+\frac{1}{1-\cos\frac{6}{7}\pi}\right)\theta_1+\left(\frac{1}{1-\cos\frac{8}{7}\pi}-\frac{1}{1-\cos\frac{2}{7}\pi}\right)(\theta_2+\theta_7)\right.$$

$$\left.+\left(\frac{1}{1-\cos\frac{8}{7}\pi}-\frac{1}{1-\cos\frac{4}{7}\pi}\right)(\theta_3+\theta_6)\right\},$$

say
$$\frac{dx_1}{dt}=a\theta_1+\beta(\theta_2+\theta_7)+\gamma(\theta_3+\theta_6)\ \dots\dots\dots\dots(122),$$

with similar equations for $x_2\dots$

Again, by equation (111),

$$r\frac{d\theta_1}{dt}=-\frac{m}{2\pi r^2}\left(4x_1+\frac{x_2}{1-\cos\frac{2}{7}\pi}+\frac{x_3}{1-\cos\frac{4}{7}\pi}+\frac{x_4}{1-\cos\frac{6}{7}\pi}\right.$$

$$\left.+\frac{x_5}{1-\cos\frac{8}{7}\pi}+\frac{x_6}{1-\cos\frac{10}{7}\pi}+\frac{x_7}{1-\cos\frac{12}{7}\pi}\right),$$

or since $x_1+x_2+x_3+\dots x_7=0,$

$$r\frac{d\theta_1}{dt}=-\frac{m}{2\pi r^2}\left\{\left(4-\frac{1}{1-\cos\frac{6}{7}\pi}\right)x_1-\left(\frac{1}{1-\cos\frac{6}{7}\pi}-\frac{1}{1-\cos\frac{2}{7}\pi}\right)(x_2+x_7)\right.$$

$$\left.-\left(\frac{1}{1-\cos\frac{8}{7}\pi}-\frac{1}{1-\cos\frac{6}{7}\pi}\right)(x_3+x_6)\right\};$$

say
$$r^2\frac{d\theta_1}{dt}=-\{ax_1-\beta(x_2+x_7)+\gamma(x_3+x_6)\}\dots(123),$$

with similar equations for $\theta_2\dots$

By means of equations (122) and (123), we get

$$\frac{d^2x_1}{dt^2}=-\{ex_1+f(x_2+x_7)+g(x_3+x_6)\},$$

where
$$e = a\alpha - 2\beta^2 + 2\beta\gamma - \gamma^2,$$
$$f = \beta(a - \alpha) + \gamma^2,$$
$$g = 2\beta\gamma + \gamma^2 - \beta^2 + \gamma(a - \alpha),$$
with similar equations for $x_2 \ldots x_6$.

Thus, if $x_1, x_2 \ldots x_7$ vary as $\epsilon^{\lambda t}$, the equation to determine λ is

$$\begin{vmatrix} \lambda^2+e, & f, & g, & 0, & 0, & g, & f \\ f, & \lambda^2+e, & f, & g, & 0, & 0, & g \\ g, & f, & \lambda^2+e, & f, & g, & 0, & 0 \\ 0, & g, & f, & \lambda^2+e, & f, & g, & 0 \\ 0, & 0, & g, & f & \lambda^2+e, & f, & g \\ g, & 0, & 0, & g, & f, & \lambda^2+e, & f \\ f, & g, & 0, & 0, & g, & f, & \lambda^2+e \end{vmatrix} = 0.$$

If $1, \omega, \omega^2, \omega^3, \omega^4, \omega^5, \omega^6$ be the seventh roots of unity, this splits up into the factors

$$\lambda^2 + e + 2f + 2g,$$
$$\{\lambda^2 + e + f(\omega + \omega^6) + g(\omega^2 + \omega^5)\}^2,$$
$$\{\lambda^2 + e + f(\omega^2 + \omega^5) + g(\omega^4 + \omega^3)\}^2,$$
$$\{\lambda^2 + e + f(\omega^3 + \omega^4) + g(\omega^6 + \omega)\}^2.$$

We proceed to calculate the numerical values of the roots

$$\alpha = 8{\cdot}52606\,\frac{m}{2\pi r}, \qquad e = 21{\cdot}70367\,\frac{m^2}{(2\pi r^2)^2},$$

$$\beta = -2{\cdot}130\,\frac{m}{2\pi r}, \qquad f = 10{\cdot}84621\,\frac{m^2}{(2\pi r^2)^2},$$

$$\gamma = -{\cdot}29192\,\frac{m}{2\pi r}, \qquad g = -1{\cdot}73332\,\frac{m^2}{(2\pi r^2)^2},$$

$$a = 3{\cdot}47394\,\frac{m}{2\pi r}.$$

Now one value of λ^2 is
$$-(e + 2\cos\tfrac{6}{7}\pi\,.\,f + 2\cos\tfrac{2}{7}\pi\,.\,g).$$

If we substitute the values of e, f, g, we shall find that this
$$= {\cdot}002\,\frac{m^2}{(2\pi r^2)^2}.$$

Thus one root of the equation in λ^2 is positive, therefore the steady motion is unstable. We therefore conclude that six is the greatest number of vortices which can be arranged at equal intervals round the circumference of the circle.

§ 54. Sir William Thomson mentions this subject in a paper in *Nature*, vol. XVIII. p. 13; in connection with some experiments made by Mr A. M. Mayer. Mr Mayer investigated experimentally the stability of various configurations of long thin magnets floating in water, and subject to the attraction of an independent fixed magnet. Sir William Thomson in the paper just mentioned, points out that if any configuration of the floating magnets form a system in stable equilibrium, the same configuration of straight columnar vortices will form a system whose steady motion is stable. Mr Mayer in his paper (*Nature*, vol. XVIII. p. 258) states that he finds the equilibrium to be stable when the floating magnets are arranged at the angular points of an equilateral triangle, a square, or a regular pentagon; but unstable for the hexagon and all polygons with a larger number of sides. This would show that the steady motion is stable for three, four, and five straight columnar vortices arranged at equal intervals round the circumference of a circle, which agrees with what we have just proved, while we have proved in addition that the steady motion of six equal vortices arranged in the same way is stable, or that seven is the smallest number of vortices which makes this way of arranging them unstable.

§ 55. To sum up the results of this section; we began by finding the motion of two vortex rings which are approximately circular, and which are linked through each other any number of times; we proved that the motion was stable when the distance between the rings was small compared with their apertures, and found the times of oscillation; we found that for each displacement there are two periods of vibration, a quick vibration, whose period is

$$\frac{2\pi}{\left(\frac{m+m'}{\pi d^2} - \frac{mm'}{m+m'}\frac{1}{4\pi a^2}(2n^2-1)\log\frac{d^2}{ee'}\right)},$$

and a slow one, whose period is

$$\frac{2\pi a}{n\sqrt{(n^2-1)}\,V},$$

where n is the order of the displacement and V the velocity of translation of the rings. We next proved that if the vortex rings are of equal strength, the condition for the possibility of motion of the kind we are considering, is that the resultant moment of momentum should be small compared with $\sqrt{\left(\frac{I^3}{m}\right)}$, where I is what Sir William Thomson calls the force resultant of the impulse, and m is the strength of the vortex.

In the case of two vortices of unequal strengths, we proved that for motion of this kind to be possible the resultant moment of momentum must have a certain definite value given by equation (105). We then proved that we might not only have 2, but 3, 4, 5, or 6 vortices twisted round each other in such a way that they all lie on the surface of an anchor ring, and are arranged in such a way that their central lines of vortex core cut any transverse section of the anchor ring in the angular points of a regular polygon inscribed in the circular transverse section. We found the times of vibration in each of these cases, and proved that the motion is unstable if seven or more vortices are arranged in this way.

PART IV.

THE VORTEX ATOM THEORY OF GASES.

§ 56. IN this part we shall consider the application to the vortex atom theory of the results we have obtained in the preceding pages. The expression we obtained in Part II. for the action of one vortex on another, would enable us to work out a dynamical theory of gases; to do this, however, would make the present essay too long, and it must form the subject of a future paper. There are, however, some results which can be obtained with very little additional calculation, and it is these results we shall consider in the following discussion.

The pressure of a gas is one of the first things a kinetic theory of gases has to explain. Sir William Thomson gives the following explanation of the pressure of a gas on the vortex atom theory (*Nature*, vol. XXIV. p. 47).

"When a vortex ring is approaching a plane, large in comparison with the dimensions of the ring, the total pressure over the surface is nil. When a ring approaches such a surface it begins to expand, so that if we consider a finite portion of the surface, the total pressure upon it due to the ring will have a finite value when the ring is close enough. In a closed cylinder, any vortex ring approaching the plane end will expand out along the surface, losing in speed as it so does, until it reaches the cylindrical boundary, along which it will crawl back on rebounding to the other end of the cylinder. As it approaches, it will therefore exert upon the plane surface a definite outward pressure whose time integral is equal to the original momentum of the vortex, and a precisely equal pressure as it leaves the surface. Hence, in the case of myriads of vortex rings bombarding such a plane surface, though no individual vortex ring leaves the surface immediately after collision, for every vortex ring that gets entangled in the condensed layer of drawn-out vortex rings another will get free, so that in the statistics of vortex impacts, the pressure exerted by a gas composed of vortex atoms is exactly the same as is given by the ordinary kinetic theory which regards the atoms as hard elastic particles."

Hence we see that just as in the ordinary solid particle theory of gases

$$p = \tfrac{2}{3}\Sigma_1 (\mathfrak{I} V'),$$

where p is the pressure, I the momentum of a vortex ring and V' its velocity, the summation being taken for all the molecules in a unit of volume of the gas, hence if v be the volume of the gas,

$$pv = \tfrac{2}{3}\Sigma (\mathfrak{I} V'),$$

where the summation is now taken for all the molecules of the gas

But by equation (9),

$$\tfrac{2}{3}\Sigma \mathfrak{I} V' = \tfrac{1}{3} T + \tfrac{1}{3}\Sigma\left(f\frac{d\mathfrak{P}}{dt} + g\frac{d\mathfrak{Q}}{dt} + h\frac{d\mathfrak{R}}{dt}\right) - \tfrac{1}{3}\rho \iint (u^2 + v^2 + w^2)\, p' ds,$$

where T is the kinetic energy, f, g, h the coordinates of the centre of a vortex ring, $\mathfrak{P}$, $\mathfrak{Q}$, $\mathfrak{R}$ the x, y, z components of the momentum of the ring, p' the perpendicular from the origin on the tangent plane to the surface containing the vortex rings. To apply this formula to gases we must calculate the value of the quantity $\Sigma\left(f\frac{d\mathfrak{P}}{dt} + g\frac{d\mathfrak{Q}}{dt} + h\frac{d\mathfrak{R}}{dt}\right)$ for any two vortex rings, say vortex (I) and vortex (II). Take the centre of vortex (I) as the origin of coordinates, the vortex (I) will contribute nothing to the term since for it f, g, h are all zero, and if a be the radius of vortex (II) l, m, n the direction cosines of its direction of motion, then $\Sigma\left(f\frac{d\mathfrak{P}}{dt} + g\frac{d\mathfrak{Q}}{dt} + h\frac{d\mathfrak{R}}{dt}\right)$ for the two vortices

$$= 2m\pi\rho a\left\{2(fl + gm + hn)\frac{da}{dt} + a\left(f\frac{dl}{dt} + g\frac{dm}{dt} + h\frac{dn}{dt}\right)\right\}.$$

If Ω be the velocity potential due to vortex (I), then if we substitute the values of $\frac{da}{dt}, \frac{dl}{dt}, \frac{dm}{dt}, \frac{dn}{dt}$ given in the equations (79) and (80), we find that the expression we are considering

$$= -2m\pi\rho a^2\left(f\frac{d}{dx} + g\frac{d}{dy} + h\frac{d}{dz}\right)\left(l\frac{d}{dx} + m\frac{d}{dy} + n\frac{d}{dz}\right)\Omega;$$

or if r be the distance between the centres of the vortex rings (I) and (II) and S the velocity perpendicular to the plane of vortex (II) due to vortex (I), this

$$= -2m\pi\rho a^2 r\frac{dS}{dr}.$$

If ϵ be the angle between the directions of motion of the vortices, and θ, θ' the angles their directions of motion make with the line joining their centres, we may easily prove that this term

$$= \frac{3mm'\pi\rho_1 a^2 a'^2}{r^3}(3\cos\theta\cos\theta' - \cos\epsilon),$$

if m' be the strength and a' the radius of vortex (I), and we see that the sum of all these terms if the vortices are not very unevenly distributed is positive, and tends to become zero.

In this investigation we have supposed that the molecules of the gas are monatomic. When the molecules are diatomic this investigation applies to that part of the term

$$\Sigma\left(f\frac{d\mathfrak{P}}{dt} + g\frac{d\mathfrak{Q}}{dt} + h\frac{d\mathfrak{R}}{dt}\right)$$

which arises from the action of one molecule on another, there will however be another part due to the action of the two atoms in a molecule on each other. To calculate this part let us for the sake of simplicity suppose that the planes of the two vortex ring atoms are parallel to each other and perpendicular to the line joining the centres of the vortex rings.

Take the centre of one of the vortex rings, say the one in the rear as the origin of coordinates, and the plane of this ring as the plane of xy; then for the ring in the rear $f = g = h = 0$, for the ring in front $f = g = 0$, $h = d'$; $\mathfrak{P} = 0$, $\mathfrak{Q} = 0$, $\mathfrak{R} = 2\pi\rho m a^2$, if a be the radius of the ring in front, d' the distance between the planes of the rings, and m the strength of either ring.

The maximum value of $\frac{d\mathfrak{R}}{dt} = 4\pi\rho m a \cdot \frac{m}{\pi d}$

(§ 37), where d is the shortest distance between the central lines of the vortex cores of the two vortex rings. Since d is small compared with a it will remain very approximately constant (§ 37); hence the greatest value of d' is d, hence $4\rho m^2 a$ is the maximum value of

$$\Sigma\left(f\frac{d\mathfrak{P}}{dt} + g\frac{d\mathfrak{Q}}{dt} + h\frac{d\mathfrak{R}}{dt}\right)$$

but $\mathfrak{J}V'$ is of the form $\rho m^2 a \log\frac{8a}{e}$ where e is the radius of the cross section of the vortex core, and $\log\frac{8a}{e}$ is very great, so that the part of the term

$$\Sigma\left(f\frac{d\mathfrak{P}}{dt} + g\frac{d\mathfrak{Q}}{dt} + h\frac{d\mathfrak{R}}{dt}\right),$$

due to the action of the two atoms in a molecule, is positive but vanishingly small compared with $\Sigma\mathfrak{I}V'$. Thus in a gas whose molecules are evenly distributed we have

$$pv = \tfrac{2}{3}\Sigma\mathfrak{I}V' = \tfrac{1}{3}T - \tfrac{1}{6}\rho\iint(u^2 + v^2 + w^2)\,p'\,dS;$$

where ρ is the density of the fluid forming the vortex rings and is not the same as the density of the gas. Since the surface is at rest the velocity of the fluid in contact with it will be small; thus the second term on the right hand will be small, if we neglect this term we have

$$pv = \tfrac{1}{3}T.$$

This is Boyle's law, while if we take into account the last term.

$$pv = \tfrac{1}{3}T - \tfrac{1}{6}\rho\iint(u^2 + v^2 + w^2)\,p'\,dS;$$

hence pv is a little less than the value given by Boyle's law, which agrees with the results of Regnault's experiments; thus the vortex atom theory explains the deviation of gases from Boyle's law. In this respect it compares favourably with the ordinary theories, for if we assume the molecules to be elastic spheres we cannot explain any deviation from Boyle's law, while if we assume that the atoms repel one another with a force varying inversely as the fifth power of the distance, the deviation ought to be the other way, *i.e.* pv ought to be greater than the value given by Boyle's law, which is contrary to the experimental results.

§ 57. According to the vortex atom theory as the temperature rises and the energy increases, the mean radius of the vortex rings will increase, but when the radius of a vortex ring is increased its velocity is diminished, and thus the mean velocity of the molecules decreases as the temperature increases; thus it differs from the ordinary kinetic theory where the mean velocity and the temperature increase together. It ought to be remarked, however, that though in the vortex atom theory the mean velocity decreases as the temperature increases, yet the mean momentum increases with the temperature.

The difference between the effects produced by a rise in temperature on the mean velocity of the molecules will probably furnish a crucial experiment between the vortex atom theory and the ordinary kinetic theory of gases, since all the laws connecting the phenomena of diffusion with the temperature can hardly be the same for the two theories. In fact, if we accept Maxwell's reasoning about the phenomenon called "thermal effusion," we can see at once an experiment which would decide between the two theories. The phenomenon is this, if we have a porous diaphragm immersed in a gas, and the gas at the two sides of the diaphragm at different temperatures, then when things have got into a steady state the pressures on the two sides of the diaphragm will be different, and Maxwell, in his paper "On Stresses in Rarified Gases," *Phil. Trans.* 1879, Part I, p. 255, gives the following reasoning to prove that, according to the ordinary theory of gases, the pressures on the two sides are proportional to the square root of the absolute temperatures of the sides. He says, "When the diameter of the hole and the thickness of the plate are both small compared with the length of the free path of the molecule, then, as Sir William Thomson has shown, any molecule which comes up to the hole on either side will be in very little danger of encountering another molecule before it has got fairly through to the other side.

"Hence, the flow of gas in either direction through the hole will take place very nearly in the same manner as if there had been a vacuum on the other side of the hole, and this whether the gas on the other side of the hole is of the same or of a different kind.

"If the gas on the two sides of the plate is of the same kind but at different temperatures, a phenomenon will take place, which we may call *thermal effusion.* The velocity of the molecules is proportional to the square root of the absolute temperature, and the quantity which passes out through the hole is proportional to this velocity and to the density. Hence, on whichever side the product of the density into the square root of the temperature is greatest, more molecules will pass from that side than from the other through the hole, and this will go on till this product is equal on both sides of the hole. Hence the condition of equilibrium is that the density must be inversely as the square root of the temperature, and since the pressure is as the product of the density into the temperature, the pressure will be directly proportional to the square root of the absolute temperature."

If we were to apply the same reasoning to the vortex atom theory, we should no longer have the velocity proportional to the square root of the absolute temperature, but to some inverse power of it, and the above reasoning would shew that if p and p' be the pressures, t and t' the temperatures on the two sides of the plate, $p/p' = (t/t')^m$ where m is a quantity greater than unity. Thus accurate investigations of the phenomenon of thermal effusion would enable us to decide between the vortex atom and the ordinary kinetic theory of gases. These experiments would, however, be difficult to make accurately, as we should have to work with such low pressures to get the mean path of the molecules long enough that the pressure of the mercury vapour in the air pump used to rarify the gas might be supposed sensibly to affect the results. In the theoretical investigation, too, the effects of the bounding surface in modifying the motion of the gas seem to have scarcely been taken sufficiently into account to make the experiment the crucial test of a theory; and it is probable that the theory of the diffusion and viscosity of gases worked out from the laws of action of two vortex rings on each other given in Part II of this essay would lead to results which would decide more easily and more clearly between the two theories.

The preceding reasoning holds only for a monatomic gas which can only increase its energy by increasing the mean radius of its vortex atoms; if however the gas be diatomic the energy will be increased if the shortest distance between the central lines of the vortex cores of the two atoms be diminished, and if the radius of the vortex atom is unaltered the velocity of translation of the molecule will be increased as well as the energy; thus for a diatomic molecule we cannot say that an increase in the energy or a rise in the temperature of the gas would necessarily be accompanied by a diminution in the mean velocity of its molecules.

§ 58. We shall now go on to apply some of the foregoing results to the case of chemical combination; in the following remarks we must be understood to refer only to bodies in the gaseous state. When two vortex rings of equal strength, with (as we shall suppose for simplicity) their planes approximately parallel to each other and approximately perpendicular to the line joining their centres, are moving in the same direction, and the circumstances are such that the hinder ring overtakes the one in front, then if, when it overtakes it, the shortest distance between the circular lines of vortex core of the rings be small compared with the radius of either ring, the rings will not separate, the shortest distance between their central lines of vortex core will remain approximately constant, and these central lines of vortex core will rotate round another circle midway between them, while this circle moves forward with a velocity of translation which is small compared with the linear velocity of the vortex rings round it. We may suppose that the union or pairing in this way of two vortex rings of different kinds is what takes place when two elements of which these vortex rings are atoms combine chemically; while, if the vortex rings are of the same kind, this process is what occurs when the atoms combine to form molecules. If two vortex rings paired in the way we have described are subjected to any disturbing influence, such as the action due to other vortex rings in their neighbourhood, their radii will be changed by different amounts; thus their velocities of translation will become different, and they will separate. We are thus led to take the view of chemical combination put forward by Clausius and Williamson, according to which the molecules of a compound gas are supposed not to always consist of the same atoms of the elementary gases, but that these atoms are continually changing partners. In order, however, that the compound gas should be something more than a mechanical mixture of the elementary gases of which it is composed, it is evidently necessary that the mean time during which an atom is paired with another of a different kind, which we shall call the paired time, should be large compared with the time during which it is alone and free from other atoms, which we shall call the free time. If we suppose that the gas is subjected to any disturbance, then this will have the effect of breaking up the molecules of the compound gas sooner than would otherwise be the case. It will thus diminish the ratio of the paired to the free time; and if the disturbance be great enough, the value of this ratio will be so much reduced that the substance will no longer exhibit the properties of a chemical compound, but those of its constituent elements: we should thus have the phenomenon of dissociation or decomposition.

We know that when two elements combine a large amount of heat is in many cases given out. We have proved in § (56) that for two vortex rings in the position of the vortex atoms of a molecule of a chemical compound $\Sigma\left(f\frac{d\mathfrak{P}}{dt}+g\frac{d\mathfrak{Q}}{dt}+h\frac{d\mathfrak{R}}{dt}\right)$ is positive; when the vortex rings are separated by a distance very great compared with the radius of either this quantity vanishes: thus we see from equation (9) that $\Sigma\mathfrak{I}V$ is increased by the combination of the atoms so that this would explain the evolution of a certain amount of heat. I do not think however that this cause would account for the enormous quantities of heat generated in some cases of chemical combination, for even these large as they are seem only to be the differences between quantities much greater than themselves. Thus for example the heat given out when hydrogen and chlorine combine to form hydrochloric acid is the difference between the heat given out when the atoms of hydrogen combine with the atoms of chlorine to form hydrochloric acid and the heat required to split up the hydrogen and chlorine molecules into their atoms. The determinations by Prof. E. Wiedemann of the heat given out when hydrogen atoms combine to form molecules, and by Prof. Thomson of the same quantity for carbon atoms seem to shew that these quantities are much greater than the quantities of heat given out in ordinary chemical combinations, and thus that these latter quantities are the differences of quantities much greater than themselves.

Whatever be the reason, the pairing of two atoms, whether of the same or different kinds, is attended by a large increase in the translatory energy. The vortex atoms however do not remain continually paired, and two atoms will only contribute to the increase in the translatory energy whilst they are paired and not when they are free, thus the whole increase in the translatory energy of a large number of molecules will depend not only on the amount of the increase contributed by any two atoms when they pair, but also on the time they remain together, and will thus depend on the ratio of the paired to the free times for the substance. The ratio of the paired to the free time plays also a very important part in determining whether chemical combination shall take place or not, and when it does take place the proportion between the amounts of the various compounds formed when more compounds than one are possible. It is clear too that the value of this ratio for the atoms of an elementary gas will have a very great effect on the chemical properties of the gas : thus if the ratio of the free to the paired times for the atoms of the gas be very small the gas will not enter readily into combination with other gases, for it will only do so to any great extent when the ratio of the free to the paired time for the compound is less than for the atoms in the molecule of the elementary gas, but if the latter be very small there is less likelihood of the ratio for the compound gas being less; thus we should expect that this ratio would be very small for the atoms of a gas like nitrogen which does not combine readily with other gases. The value of the ratio would afford a very convenient measure for the affinity of the constituents of a compound for each other. It is also conceivable that this ratio might affect the physical properties of a gas, and in a paper in the *Philosophical Magazine* for June 1883 I suggested that differences in the value of this ratio might account for the differences in the dielectric strengths of gases.

Two vortex rings will not remain long together unless the shortest distance between the central lines of their vortex cores is small compared with the radius of either of the rings; now as the vortex rings approach each other they alter in size, the one in front expands and the one in the rear contracts. If the rings are to remain together their radii must become nearly equal as they approach each other and their planes become nearly coincident: it is evident however that for this to happen the radii of the rings before they pair must lie within certain limits. The energy of the gas however, and therefore the temperature depend upon the mean radius of the vortex rings which form the atoms of the gas, and conversely the mean radius of the vortex atoms is a function of the temperature, and if the mean radius is between certain limits the temperature must also be between limits, thus unless the temperature is between certain limits the atoms would not remain long together after they paired and so chemical combination would not take place; this reasoning would indicate that chemical combination could only occur between certain limits of temperature, and this seems to be the case in at any rate a great many cases of chemical combination.

The following reasoning will explain how it is that the compound after it is formed can exist at temperatures at which the elements of which it is composed could not combine. When the elements have once combined the molecules of the compound will settle down so that the radii of their vortex atoms will be distributed according to a definite law, and a large proportion of the vortex atoms will have their radii between comparatively narrow limits, just as in the ordinary theory of gases Maxwell's law gives the distribution of velocity. Now suppose that a molecule of a compound of the elements A and B is subjected to any disturbance tending to change the radii of the atoms; though the difference in the changes in the radii may be sufficient to cause the atoms to separate, yet since the atoms were close together when they were disturbed the difference in the changes must be small, and since the motion is reversible the atom A would only have to suffer a slight change to be able to combine again with a vortex ring like B, or it could combine at once with a vortex ring differing only slightly in radius from B; thus A will have plenty of chances of recombination with the B atoms and will be in a totally different position with regard to them from that in which it would have been if it had not previously been in combination with a B atom.

Let us now suppose that two vortex rings of approximately equal radius but of different strengths come close together in such a way that their planes are approximately parallel and perpendicular to the line joining their centres, then we can see, as in the analogous case of linked vortices, that the motion will be of the following kind. Let m and m' be the strengths of the two vortices, a the mean radius of either, and d the shortest distance between their central lines, e and e' the radii of the cross sections of the two vortex rings. If we imagine a circle between the central lines of the two vortex rings dividing the distance between the vortices inversely as the strengths of the vortices, the two vortex rings will rotate round this circle with an angular velocity $m+m'/d\pi^2$ remaining at an approximately constant distance d apart, while the circle itself will move with a comparatively slow motion of translation perpendicular to its own plane.

The mean velocity of the vortex of strength m

$$=\frac{m}{2\pi a}\left(\log\frac{8a}{e}-1\right)+\frac{m'}{2\pi a}\left(\log\frac{8a}{d}-1\right);$$

the mean velocity of the vortex of strength m'

$$=\frac{m'}{2\pi a}\left(\log\frac{8a}{e'}-1\right)+\frac{m}{2\pi a}\left(\log\frac{8a}{d}-1\right).$$

Now if the two vortices are to remain together, their mean velocities must be equal; therefore

$$\frac{m}{2\pi a}\log\frac{8a}{e}+\frac{m'}{2\pi a}\log\frac{8a}{d}=\frac{m'}{2\pi a}\log\frac{8a}{e'}+\frac{m}{2\pi a}\log\frac{8a}{d}.$$

Now suppose a and d become, through some external influence, $a+\delta a$ and $d+\delta d$, then the change in the mean velocity of the vortex of strength m is, if V be the original mean velocity,

$$\left(-\frac{V}{a}+\frac{3m+2m'}{4\pi a^2}\right)\delta a-\frac{m}{2\pi a}\cdot\frac{\delta d}{d};$$

and if we interchange m and m' in this formula, we shall get the change in the mean velocity of the vortex ring whose strength is m'. Now if the two vortex rings are to remain together for a time long compared with the mean interval between two collisions, in spite of all the vicissitudes they will meet with when moving about in an enclosure containing a great number of moving molecules, the mean velocities of the two vortex rings must always remain equal; thus the changes in the mean velocities of the rings must be equal for all values of δa and δd, so that the coefficients of δa and δd must be equal in the two expressions for the changes in the mean velocities; for this to be the case we see that m must equal m'. Hence, we conclude that if two vortex rings are to remain for long together when subject to disturbing influences, they must be of equal strength. We can extend this result to the case when we have more than two vortices close together; however many vortices there are, if they are to remain together for any considerable time they must be of equal strength.

§ 59. We shall often have occasion to speak of vortex rings arranged in the way discussed in § 43, *i.e.* so that those portions of the central lines of vortex core of the several vortex rings which are closest together are always approximately parallel, and so that a plane perpendicular to their central lines at any point cuts them in the angular points of a regular polygon. We proved in Part III that if the vortices are of equal strengths, and not more than six in number, they will be in stable steady motion; it is not necessary for the truth of this proposition that each vortex ring should be single; the proposition will be true if the vortex rings are composite, provided the distances between their components are small compared with the sides of the polygon, at the angular points of which the vortices are situated, and that the sum of the strengths of the components is the same as the strength of the single vortex ring, which they are supposed to replace. We shall speak of the systems of vortices placed at the angular points of the polygon as the primaries, and the component vortex rings of these primaries as the secondaries of the system; and when we speak of a system consisting of three, four, five, or six primaries, we shall suppose, unless we expressly state the contrary, that they are arranged in the way just described.

We may imagine the way in which these vortex rings are linked through each other by supposing that we take a cylindrical rod and describe on its surface a screw with n threads; let us first suppose that there are an exact number of turns of each thread on the rod, bend the rod into a circle and join the ends, then each of the n threads of the screw will represent the central line of the vortex core of one of the n equal linked vortices; next suppose that the threads make m/n turns in the length of the rod where m is an integer not divisible by n, then if we bend the rod as before and join the ends, the threads of the screw will form an endless thread with n loops, and for the present purpose the properties of a vortex ring whose core is of this kind will be similar to those of one where the n threads are distinct, so that we may suppose the core of the vorticity which forms the atom to be arranged, in either of these ways, and we shall speak of it as an atom with n links; thus the links may be separate or run one into the other forming an endless chain.

Now let us suppose that the atoms of the different chemical elements are made up of vortex rings all of the same strength, but that some of these elements consist of only one of these rings, others of two of the rings linked together, or else of a continuous curve with two loops, others of three, and so on; but our investigation at the end of Part III shews that no element can consist of more than six of these rings if they are arranged in the symmetrical way there described.

Then if any of these atoms combine so as to form a permanent combination, the strengths of all the primaries in the system formed by the combination must be equal. Thus an atom of an element may combine with another atom of the same kind to form a molecule of the substance consisting of two atoms. Again, three of these atoms may combine and form a system consisting of three primary elements, but the chance of their doing this is small compared with the chance of two pairing, so that the number of systems of this kind will be small compared with the number of the systems consisting of only two atoms. We might have systems consisting of four atoms, but the number would be small compared with the number of systems that consist of three atoms, and so on. We could not have a system consisting of more than six primaries if arranged in the way supposed, but though this seems the most natural way of arranging the atoms, we must not be understood to assert that this is the only way, and in special cases the atoms may be arranged differently, and then we might have systems consisting of more than six primaries. Now, suppose that an atom of one element is to combine with an atom of another. Suppose to fix our ideas, that the atom consisting of two vortex rings linked together is to combine with an atom consisting of one vortex ring, then since for stability of connection, the strength of all the primaries which form the components of the compound system must be equal; the atom consisting of two links must unite with molecules containing two atoms of the one with one link. If the atoms are made to combine directly, the chance that they form the simplest combination is almost infinitely greater than the chance of any more complex combination, so that the number of the simplest compound systems will be almost infinitely greater than the number of any more complex compound system. Thus the compound formed will be the simplest combination, consisting of one of the atoms, which consist of two vortex rings linked together, with two of the atoms consisting of only one vortex ring. Similarly, if an atom consisting of three vortex rings linked together were to combine directly with atoms consisting of only one vortex ring, the compound formed would consist of one of the three linked atoms with three of the others, and so on for the combination of atoms formed by any number of vortex rings linked together. This suggests, that the atoms of the elements called by the chemists monads, dyads, triads, tetrads, and so on, consist of one, two, three, four, &c., vortex rings linked together, for then we should know that a dyad could not combine with less than two atoms of a monad to form a stable compound, a triad with less than three, and so on, which is just the definition of the terms monad, dyad, &c.

Thus each vortex ring in the atom would correspond to a unit of affinity in the chemical theory of quantivalence. If we regard the vortex rings in those atoms consisting of more vortex rings than one as linked together in the most symmetrical way, then no element could have an atom consisting of more than six vortex rings at the most, so that no single atom would be capable of uniting with more than six atoms of another element so as to form a stable compound. This agrees with chemical facts, as Lothar Meyer in his *Modernen Theorien der Chemie*, 4th Edition, p. 196, states that no compound consisting of more than six atoms of one element combined with only one of another is known to exist in the gaseous state, and that a gaseous compound of tungsten, consisting of six atoms of chlorine united to one of tungsten does exist.

Though in direct combination, the simplest compound is the one that would naturally be formed; yet other compounds are possible, and under other circumstances might be formed; thus one atom of a dyad might unite not only with two atoms of a monad, but also with four atoms of a monad, the four atoms of the monad splitting up into two groups of two each; thus the one atom of the dyad and the two groups of two monads would form three primaries, which when arranged in the way described above, would be in stable steady motion; again, we might have two atoms of the dyad and two of the monad forming again a system with three primaries, or one atom of the dyad might unit with one atom of another dyad, forming a system with two primaries, or with two atoms of another dyad forming a system with three primaries, and so on. These remarks may be illustrated by means of the following gaseous compounds of sulphur and mercury.

Thus we have the compounds:

$$H_2S,\ SO_2,\ SO_3,$$
$$Hg.Cl_2,\ Hg_2Cl_2.$$

In fact, all that is necessary for the existence of any compound from this point of view is, that its constituents should be capable of division into primaries of equal strength, and if these are to be arranged in the simplest and most symmetrical way, that there should not be more than six of them.

§ 60. Looking at chemical combinations from this point of view, we should expect to find that such compounds as hydrochloric acid, where one atom of hydrogen has only to meet with one atom of chlorine; or water where an atom of oxygen has only to meet with two atoms or a molecule of hydrogen, would be much more easily and quickly formed by direct combination, than a compound such as ammonia gas, to form which, an atom of nitrogen has to find itself close to three atoms of hydrogen at once; and it is, I believe, the case in direct combination, that simple compounds are formed more quickly than complex ones.

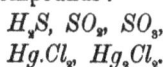

We shall call the ratio of the number of links in the atom of an element to the number in the atom of hydrogen, the valency of the element. To determine this quantity with any degree of certainty, we require to know the accurate composition of a large number of the gaseous compounds of the element; thus only those compounds whose vapour-density is known afford us any assistance, as it would make a great difference, for example, in the valency of nitrogen, if the molecule of ammonia could be represented by the formula N_2H_6 instead of NH_3, and differences of this kind can only be determined by vapour-density determinations; so that in the following discussion of the valency of the elements, too much importance must not be attached to the result for any element, when the vapour densities of only a few of its compounds have been determined. The determination of a single vapour-density will enable us to assign a superior limit to the valency of the elements in the compound, but it may require a great many vapour-density determinations to enable us to assign a lower limit to the valency of the same element.

The compounds HCl, HI, HBr, HF, Tl Cl. shew that the atoms of chlorine, iodine, fluorine, and thallium have the same number of links as the atom of hydrogen, or that the valency of each of these elements is unity. From the compound H_2O we infer that the atom of oxygen consists of twice as many links as the atom of hydrogen, though as far as this compound goes there is nothing to shew that the atom of oxygen does not consist of the same number of links as the atom of hydrogen, in this case however we should have to look upon the molecule of water as a system with the three primaries H – H – O; it is however preferable to take the simpler view that the water molecule is a system with the two primaries HH – O, and suppose that the valency of oxygen is two: the composition of all the compounds of oxygen may be explained on this supposition, and there are other considerations which lead us to endeavour to reduce the number of primaries in the molecule of a compound to as few as possible. Regarding oxygen then as a dyad, the molecule of hydrogen peroxide consists of the three primaries $H_2 - O - O$.

The compounds H_2S, H_2Se, Pb Cl_2, Cd Br_2, Te H_2, indicate that the atoms of sulphur, selenium, tellurium, lead and cadmium have twice as many links as the atom of hydrogen. The compound CO shews that the atom of carbon has the same number of links as the atom of oxygen, or twice as many as the atom of hydrogen; the molecules of carbonic acid and marsh gas have each three primaries represented by C – O – O and $C - H_2 - H_2$ respectively. Carbon is usually regarded as a tetrad, and we should therefore have expected its atom to have four times as many links as the atom of hydrogen; the compound CO shews however that if the view we have taken be correct, the carbon atom must have only twice as many links as the hydrogen atom: this view is supported by the composition of acetylene C_2H_2; if the valency of carbon atom be two, the molecule may be divided into the three primaries $C - C - H_2$, but if the valency of carbon were four, the molecule of acetylene could not be divided into primaries of equal strength, so that according to our view, its constitution is impossible on this supposition.

The sulphur compounds afford good examples of molecules containing various numbers of primaries, thus we have H_2S with two primaries $H_2 - S$; SO_2 with three primaries S – O – O and SO_3 with four primaries S – O – O – O.

It is difficult to determine from the composition of the mercury compounds as given in the chemical text-books whether the atom of mercury has the same number of links as the atom of hydrogen or twice that number; according to Lothar Meyer the composition of calomel is Hg Cl, in most of the other text-books it is given as $Hg_2\ Cl_2$; if Lothar Meyer's supposition be correct then the mercury atom has as many links as the hydrogen atom and the molecule of calomel consists of the two primaries Hg – Cl while the molecule of corrosive sublimate consists of the three primaries Hg – Cl – Cl; if however the composition of calomel is $Hg_2\,Cl_2$ then the mercury atom probably has twice as many links as the hydrogen atom and the molecule of calomel consists of the three primaries Hg – Hg – Cl_2 while the molecule of corrosive sublimate consists of the two primaries Hg – Cl_2.

The following reasons lead us to suppose that the atom of phosphorus has the same number of links as the atom of hydrogen; the composition of phosphoretted hydrogen PH_3 shews that the atom of phosphorus must either have the same number of links as the hydrogen atom in which case the molecule consists of four primaries, or it must have three times as many in which case the molecule of phosphoretted hydrogen will have two primaries; the compound PH_5 however shews that the phosphorus atom has either the same number of links as the hydrogen atom or five times as many; hence we see that the phosphorus atom must have the same number of links as the hydrogen atom. The resemblance between the properties of arsenic and phosphorus would lead us to conclude that the atom of arsenic had the same number of links as the atom of hydrogen, and the constitution of its compounds could be explained on this supposition; there is nothing to shew from its simpler inorganic compounds that the arsenic atom has not three times as many links as the hydrogen atom; the composition of the chloride of cacodyl As Cl $C_2\,H_6$ shews however that this is not the case and the atom of arsenic like that of phosphorus must have the same number of links as the hydrogen atom.

The compounds of nitrogen present great difficulties when considered from this point of view; the composition of ammonia NH_3 requires us to suppose either that the nitrogen atom has three times as many links as the hydrogen atom, in which case the molecule of ammonia would consist of the two primaries N – H_3, or that the nitrogen atom has the same number of links as the hydrogen atom and then the molecule of ammonia would consist of the four primaries N – H – H – H; the composition of nitric oxide NO however compels us to suppose that the atom of nitrogen has the same number of links as the atom of oxygen or twice as many as the atom of hydrogen, and these suppositions are inconsistent. It is however conceivable that an atom might go through a process that would cause it to act like one with twice as many links. To illustrate this take a single circular ring and pull the opposite sides so that they cross at the centre of the ring, forming a figure of eight, then bend one half of the figure of eight over the other half, the continuous ring will now form two circles whose planes are nearly coincident. If the circular ring represented a line of vortex core the duplicated ring would behave like one with twice as many links as the original ring. Thus if we look upon the atom of nitrogen as consisting of the same number of links as the atom of hydrogen we can explain the constitution of the compounds NH_3, N_2O, N_2O_3, C_2N_2, HCN, CNH_5 &c., but in the compounds NO, NO_2 we should have to suppose that the atom was duplicated in the manner described above.

The following table shews the valency of those elements which form gaseous compound of known vapour density, though as we said before when we know the vapour density of only a few of the compounds of an element the value given in the table must not be looked on as anything more than an upper limit to the value of the valency of the element.

Univalent Elements.

Arsenic.	Mercury ?
Bromine.	Nitrogen.
Chlorine.	Phosphorus.
Fluorine.	Potassium.
Hydrogen.	Rubidium.
Iodine.	Thallium.

Divalent Elements.

Cadmium.	Mercury ?
Carbon.	Oxygen.
Chromium.	Selenium.
Copper.	Sulphur.
Lead.	Tellurium.
Manganese.	Zinc.

Trivalent Elements.

Aluminium.	Bismuth.
Antimony.	Boron.

Indium.

Quadrivalent Elements.

Silicon.	Tin.

§ 61. According to the view we have taken, atomicity corresponds to complexity of atomic arrangement; and the elements of high atomicity consist of more vortex rings than those whose atomicity is low; thus high atomicity corresponds to complicated atomic arrangement, and we should expect to find the spectra of bodies of low atomicity much simpler than those of high. This seems to be the case, for we find that the spectra of Sodium, Potassium, Lithium, Hydrogen, Chlorine which are all monad elements, consist of comparatively few lines.